W9-ADN-632

# Uniform Convexity, Hyperbolic Geometry, and Nonexpansive Mappings

## MONOGRAPHS AND TEXTBOOKS IN PURE AND APPLIED MATHEMATICS

1. *K. Yano,* Integral Formulas in Riemannian Geometry (1970) *(out of print)*
2. *S. Kobayashi,* Hyperbolic Manifolds and Holomorphic Mappings (1970) *(out of print)*
3. *V. S. Vladimirov,* Equations of Mathematical Physics (A. Jeffrey, editor; A. Littlewood, translator) (1970) *(out of print)*
4. *B. N. Pshenichnyi,* Necessary Conditions for an Extremum (L. Neustadt, translation editor; K. Makowski, translator) (1971)
5. *L. Narici, E. Beckenstein, and G. Bachman,* Functional Analysis and Valuation Theory (1971)
6. *D. S. Passman,* Infinite Group Rings (1971)
7. *L. Dornhoff,* Group Representation Theory (in two parts). Part A: Ordinary Representation Theory. Part B: Modular Representation Theory (1971, 1972)
8. *W. Boothby and G. L. Weiss (eds.),* Symmetric Spaces: Short Courses Presented at Washington University (1972)
9. *Y. Matsushima,* Differentiable Manifolds (E. T. Kobayashi, translator) (1972)
10. *L. E. Ward, Jr.,* Topology: An Outline for a First Course (1972) *(out of print)*
11. *A. Babakhanian,* Cohomological Methods in Group Theory (1972)
12. *R. Gilmer,* Multiplicative Ideal Theory (1972)
13. *J. Yeh,* Stochastic Processes and the Wiener Integral (1973) *(out of print)*
14. *J. Barros-Neto,* Introduction to the Theory of Distributions (1973) *(out of print)*
15. *R. Larsen,* Functional Analysis: An Introduction (1973) *(out of print)*
16. *K. Yano and S. Ishihara,* Tangent and Cotangent Bundles: Differential Geometry (1973) *(out of print)*
17. *C. Procesi,* Rings with Polynomial Identities (1973)
18. *R. Hermann,* Geometry, Physics, and Systems (1973)
19. *N. R. Wallach,* Harmonic Analysis on Homogeneous Spaces (1973) *(out of print)*
20. *J. Dieudonné,* Introduction to the Theory of Formal Groups (1973)
21. *I. Vaisman,* Cohomology and Differential Forms (1973)
22. *B.-Y. Chen,* Geometry of Submanifolds (1973)
23. *M. Marcus,* Finite Dimensional Multilinear Algebra (in two parts) (1973, 1975)
24. *R. Larsen,* Banach Algebras: An Introduction (1973)
25. *R. O. Kujala and A. L. Vitter (eds.),* Value Distribution Theory: Part A; Part B. Deficit and Bezout Estimates by Wilhelm Stoll (1973)
26. *K. B. Stolarsky,* Algebraic Numbers and Diophantine Approximation (1974)
27. *A. R. Magid,* The Separable Galois Theory of Commutative Rings (1974)
28. *B. R. McDonald,* Finite Rings with Identity (1974)

29. *J. Satake,* Linear Algebra (S. Koh, T. Akiba, and S. Ihara, translators) (1975)
30. *J. S. Golan,* Localization of Noncommutative Rings (1975)
31. *G. Klambauer,* Mathematical Analysis (1975)
32. *M. K. Agoston,* Algebraic Topology: A First Course (1976)
33. *K. R. Goodearl,* Ring Theory: Nonsingular Rings and Modules (1976)
34. *L. E. Mansfield,* Linear Algebra with Geometric Applications: Selected Topics (1976)
35. *N. J. Pullman,* Matrix Theory and Its Applications (1976)
36. *B. R. McDonald,* Geometric Algebra Over Local Rings (1976)
37. *C. W. Groetsch,* Generalized Inverses of Linear Operators: Representation and Approximation (1977)
38. *J. E. Kuczkowski and J. L. Gersting,* Abstract Algebra: A First Look (1977)
39. *C. O. Christenson and W. L. Voxman,* Aspects of Topology (1977)
40. *M. Nagata,* Field Theory (1977)
41. *R. L. Long,* Algebraic Number Theory (1977)
42. *W. F. Pfeffer,* Integrals and Measures (1977)
43. *R. L. Wheeden and A. Zygmund,* Measure and Integral: An Introduction to Real Analysis (1977)
44. *J. H. Curtiss,* Introduction to Functions of a Complex Variable (1978)
45. *K. Hrbacek and T. Jech,* Introduction to Set Theory (1978) *(out of print)*
46. *W. S. Massey,* Homology and Cohomology Theory (1978)
47. *M. Marcus,* Introduction to Modern Algebra (1978)
48. *E. C. Young,* Vector and Tensor Analysis (1978)
49. *S. B. Nadler, Jr.,* Hyperspaces of Sets (1978)
50. *S. K. Sehgal,* Topics in Group Rings (1978)
51. *A. C. M. van Rooij,* Non-Archimedean Functional Analysis (1978)
52. *L. Corwin and R. Szczarba,* Calculus in Vector Spaces (1979)
53. *C. Sadosky,* Interpolation of Operators and Singular Integrals: An Introduction to Harmonic Analysis (1979)
54. *J. Cronin,* Differential Equations: Introduction and Quantitative Theory (1980)
55. *C. W. Groetsch,* Elements of Applicable Functional Analysis (1980)
56. *I. Vaisman,* Foundations of Three-Dimensional Euclidean Geometry (1980)
57. *H. I. Freedman,* Deterministic Mathematical Models in Population Ecology (1980)
58. *S. B. Chae,* Lebesgue Integration (1980)
59. *C. S. Rees, S. M. Shah, and Č. V. Stanojević,* Theory and Applications of Fourier Analysis (1981)
60. *L. Nachbin,* Introduction to Functional Analysis: Banach Spaces and Differential Calculus (R. M. Aron, translator) (1981)
61. *G. Orzech and M. Orzech,* Plane Algebraic Curves: An Introduction Via Valuations (1981)

62. *R. Johnsonbaugh and W. E. Pfaffenberger,* Foundations of Mathematical Analysis (1981)
63. *W. L. Voxman and R. H. Goetschel,* Advanced Calculus: An Introduction to Modern Analysis (1981)
64. *L. J. Corwin and R. H. Szczarba,* Multivariable Calculus (1982)
65. *V. I. Istrăţescu,* Introduction to Linear Operator Theory (1981)
66. *R. D. Järvinen,* Finite and Infinite Dimensional Linear Spaces: A Comparative Study in Algebraic and Analytic Settings (1981)
67. *J. K. Beem and P. E. Ehrlich,* Global Lorentzian Geometry (1981)
68. *D. L. Armacost,* The Structure of Locally Compact Abelian Groups (1981)
69. *J. W. Brewer and M. K. Smith, eds.,* Emmy Noether: A Tribute to Her Life and Work (1981)
70. *K. H. Kim,* Boolean Matrix Theory and Applications (1982)
71. *T. W. Wieting,* The Mathematical Theory of Chromatic Plane Ornaments (1982)
72. *D. B. Gauld,* Differential Topology: An Introduction (1982)
73. *R. L. Faber,* Foundations of Euclidean and Non-Euclidean Geometry (1983)
74. *M. Carmeli,* Statistical Theory and Random Matrices (1983)
75. *J. H. Carruth, J. A. Hildebrant, and R. J. Koch,* The Theory of Topological Semigroups (1983)
76. *R. L. Faber,* Differential Geometry and Relativity Theory: An Introduction (1983)
77. *S. Barnett,* Polynomials and Linear Control Systems (1983)
78. *G. Karpilovsky,* Commutative Group Algebras (1983)
79. *F. Van Oystaeyen and A. Verschoren,* Relative Invariants of Rings: The Commutative Theory (1983)
80. *I. Vaisman,* A First Course in Differential Geometry (1984)
81. *G. W. Swan,* Applications of Optimal Control Theory in Biomedicine (1984)
82. *T. Petrie and J. D. Randall,* Transformation Groups on Manifolds (1984)
83. *K. Goebel and S. Reich,* Uniform Convexity, Hyperbolic Geometry, and Nonexpansive Mappings (1984)

*Other Volumes in Preparation*

# Uniform Convexity, Hyperbolic Geometry, and Nonexpansive Mappings

Kazimierz Goebel

*Institute of Mathematics*
*Maria Curie-Skłodowska University*
*Lublin, Poland*

Simeon Reich

*Department of Mathematics*
*The University of Southern California*
*Los Angeles, California*

**MARCEL DEKKER, INC. New York and Basel**

Library of Congress Cataloging in Publication Data

Goebel, Kazimierz, [date]
Uniform convexity, hyperbolic geometry, and non-expansive mappings.

(Monographs and textbooks in pure and applied mathematics ; v. 83)
1. Banach spaces. 2. Holomorphic mappings.
3. Nonexpansive mappings. 4. Geometry, Hyperbolic.
I. Reich, Simeon. II. Title. III. Series.
QA322.2.G64 1983 515.7'32 84-4978
ISBN 0-8247-7223-7



MARCEL DEKKER, INC.
270 Madison Avenue, New York, New York 10016

Current printing (last digit):
10 9 8 7 6 5 4 3 2 1

PRINTED IN THE UNITED STATES OF AMERICA

# PREFACE

In recent years, rapid developments have occurred in two seemingly unrelated theories. The settings of both theories are (complex, infinite-dimensional) Banach spaces, but one deals with nonexpansive mappings, while the other deals with holomorphic mappings. The main purpose of these notes (which originate from lectures given at the Nonlinear Functional Analysis Seminar at U.S.C. during the first author's visit there) is to expose the surprising connections between these two theories.

The theory of nonexpansive mappings (that is, those mappings with Lipschitz constant 1) on convex subsets of Banach spaces is a recent branch of nonlinear functional analysis. It has flourished during the last twenty years with many papers, results, and still unsolved problems. It is intimately connected with differential equations and with the geometry of Banach spaces. The theory of holomorphic mappings on domains in complex Banach spaces is rooted in classical complex analysis and hyperbolic geometry. However, the theories of holomorphic functions of several complex variables and of infinite-dimensional holomorphy are quite young.

One link between these two theories is the fact that holomorphic mappings are nonexpansive with respect to certain pseudometrics. A second, deeper link is the concept of uniform convexity which, as it turns out, plays a central role in both theories.

There are relatively few books which try to bridge the gap between two different areas of mathematics. This is our intention in the present text. Therefore, our exposition is as elementary as possible. Thus, we use only metric methods and avoid using more advanced analytic methods and differential geometry. Consequently, these notes should be accessible to readers with only a modest background in functional analysis and complex analysis.

On the other hand, we hope these lecture notes are of interest to specialists of both camps. (Especially to those who are interested in finding out what the other camp has been doing.) We have included many new and recent results, as well as open problems.

The text is divided into three chapters. Although in the first chapter we are mainly interested in Hilbert space (the infinite-dimensional analog of Euclidean geometry), we actually consider the more general Banach space case. This is because the Banach space case, in addition to its intrinsic interest, illuminates and provides insight into the Hilbert space case. We begin the second chapter with a discussion of the Poincaré metric on the open unit disc in the plane. The main topic in this chapter is the Hilbert ball, the infinite-dimensional analog of Lobachevsky's hyperbolic geometry, but we study other domains too. We end this text with a short chapter on the Hilbert sphere (the infinite-dimensional analog of the Riemann sphere). In all three chapters, we use uniform (metric) convexity to study nonexpansive mappings with special emphasis on fixed point theory. The Table of Contents gives more information on the topics covered in each one of the chapters.

Our text is not intended to be a comprehensive monograph. Instead, we view it as an introductory text to two exciting fields. We hope it leads to more work and to a better understanding of both.

For other aspects of recent work in hyperbolic geometry and non-Euclidean functional analysis we refer the reader to the recent articles of Helton, Milnor, and Thurston.

The second author gratefully acknowledges the support given by the National Science Foundation and the USC Faculty Research and Innovation Fund.

Special thanks go to Leon Lemons and Cynthia Summerville for typing the manuscript.

Kazimierz Goebel

Simeon Reich

# CONTENTS

CHAPTER 2 HYPERBOLIC GEOMETRY

# Uniform Convexity, Hyperbolic Geometry, and Nonexpansive Mappings

# 1

# BANACH SPACES

## 1. BASIC FACTS

In this section we recall several basic properties of self-mappings of metric spaces. Let $(M,d)$ be a metric space. A mapping $T : M \to M$ is said to be Lipschitzian if there exists a non-negative number $k$ such that

$$d(Tx,Ty) \leq kd(x,y)$$

for all $x$ and $y$ in $M$. The smallest such $k$ is called the Lipschitz constant of $T$ and will be denoted by $L(T)$.

The mapping $T$ is said to be a strict contraction if $L(T) < 1$. It is said to be nonexpansive if $L(T) \leq 1$. If $S$ is another self-mapping of $M$ and $S \circ T$ is the composition of $T$ and $S$, then

$$L(S \circ T) \leq L(S) \cdot L(T).$$

In particular, if $T^n$ denotes the n-th iterate of $T$ ($T$ composed with itself n times), then $L(T^n) \leq [L(T)]^n$. A point $y$ in $M$ is said to be a fixed point of $T$ if $Ty = y$.

The following result (Banach's fixed point theorem) is the simplest and perhaps the most useful in fixed point theory.

<u>Theorem 1.1.</u> Let T be a self-mapping of a complete metric space M. If T is a strict contraction, then it has a unique fixed point y and $\lim_{n\to\infty} T^n x = y$ for all x in M.

<u>Proof.</u> Let $k < 1$ be the Lipschitz constant of T, and let x be a point in M. Denoting the metric of M by d, we have for any n and p,

$$d(T^n x, T^{n+p} x) \leq \sum_{i=0}^{p-1} d(T^{n+i} x, T^{n+i+1} x)$$
$$\leq \sum_{i=0}^{p-1} k^{n+i} d(x,Tx) \leq \frac{k^n}{1-k} d(x,Tx).$$

This shows that $\{T^n x\}$ is a Cauchy sequence. Therefore $\lim_{n\to\infty} T^n x = y$ exists. Since

$$Ty = \lim_{n\to\infty} T(T^n x) = \lim_{n\to\infty} T^{n+1} x = y,$$

y is indeed a fixed point of T. If z is another fixed point, then $d(y,z) = d(Ty,Tz) \leq kd(y,z)$. Hence $d(y,z) = 0$ and $y = z$.

The existence of a fixed point in Theorem 1.1 may be proved in another simple way. Let $a = \inf\{d(x,Tx) : x \in M\}$. If $\varepsilon > 0$ and $d(x,Tx) < a + \varepsilon$, then $a \leq d(Tx,T^2x) \leq k\, d(x,Tx) < ka + \varepsilon$. Hence $a = 0$. Now consider the sets $M_\varepsilon = \{x \in M: d(x,Tx) \leq \varepsilon\}$. Each one of these sets in nonempty and closed. If both x and y belong to $M_\varepsilon$, then

$$d(x,y) \leq d(x,Tx) + d(Tx,Ty) + d(Ty,y) \leq 2\varepsilon + kd(x,y).$$

Denoting the diameter of $M_\varepsilon$ by diam $(M_\varepsilon)$, we thus see that diam $(M_\varepsilon) \leq 2\varepsilon/(1-k) \underset{\varepsilon\to 0}{\to} 0$. Since M is complete, the intersection of $\{M_\varepsilon : \varepsilon > 0\}$ must therefore consist of exactly one point. It is clear that this point must be the unique fixed point of T.

These two ideas of iteration and intersection will reappear time and again throughout the text.

What happens to Banach's fixed point theorem if $L(T) = 1$? We begin our discussion of this question with several simple observations.

Even an isometry of a complete metric space may fail to have a fixed point: the shift of the real axis R defined by $Tx = x+1$ is the simplest example. Rotations of the unit circle in the complex plane C show that this is true even for compact spaces.

On the other hand, positive results can be obtained for those mappings $T : M \to M$ that satisfy

$$d(Tx,Ty) < d(x,y) \text{ for } x \neq y.$$

Such mappings are said to be contractive.

In fact, let M be compact and $T : M \to M$ contractive. Let the continuous function $f: M \to R$ defined by $f(x) = d(x,Tx)$ attain its minimum m over M at $y \in M$. If $y \neq Ty$, then

$$m \leq d(Ty,T^2y) < d(y,Ty) = m.$$

Hence $y = Ty$ is the unique fixed point of T.

If M is not compact, then a contractive mapping need not have a fixed point. To see this, consider the mappings $T: R \to R$ and $S : [1,\infty) \to [1,\infty)$ defined by $Tx = \log(1+e^x)$ and $Sx = x + 1/x$ respectively. We do have, however, the following result, due to M. Edelstein [30].

Theorem 1.2. Let M be a metric space and $T : M \to M$ a contractive mapping. If there is a point z in M such that a subsequence of $\{T^nz\}$ converges to y, then y is the unique fixed point of T, and $\lim_{n\to\infty} T^nz = y$.

Proof. The sequence $\{d(T^nz,T^{n+1}z)\}$ decreases to a limit which equals both $d(y,Ty)$ and $d(Ty,T^2y)$. If $y \neq Ty$, then

$$d(Ty,T^2y) < d(y,Ty).$$

Hence y must be a fixed point of T. Therefore the sequence $\{d(T^nz,y)\}$ also decreases to a limit, which must equal 0.

Let us say that a metric space M is finitely compact if each bounded closed subset of M is compact. Theorem 1.2 provides us with a useful fact concerning such spaces.

Theorem 1.3. Let $(M,d)$ be a finitely compact metric space, $T : M \to M$ a contractive mapping, and w a point in M. If there is a point z in M such that

$$\liminf_{n\to\infty} d(w,T^nz) < \infty,$$

then T has a unique fixed point y, and $\lim_{n\to\infty} T^nx = y$ for all x in M.

We remark in passing that recent applications of Theorem 1.2 can be found in the paper by J. Reinermann, G.H. Seifert, and V. Stallbohm [90].

Turning to Banach spaces, we first observe the following simple but important fact.

Proposition 1.4. Let C be a bounded closed convex subset of a Banach space $(X,|\cdot|)$. If $T : C \to C$ is nonexpansive, then $\inf\{|x - Tx| : x \in C\} = 0$.

Proof. Fix a point y in C, and consider the mappings $T_t : C \to C$, $0 \leq t < 1$, defined by $T_tx = (1-t)y + tTx$ for x in C. Since $L(T_t) \leq t$, each $T_t$ has a unique fixed point $x_t$. Now we have

$$|x_t-Tx_t| = |(1-t)y + tTx_t - Tx_t| = (1-t)|y - Tx_t|$$

$$\leq (1-t) \text{ diam } (C),$$

and the result follows by letting $t \to 1$.

If $C$ is also compact, then Proposition 1.4 shows that $T$ must have a fixed point. (This fact is also a consequence of Schauder's fixed point theorem.) Such a fixed point is no longer unique.

If $C$ is not compact, then $T$ need not have a fixed point.

Example 1.5. Let $X = C[0,1]$ and

$$C = \{x \in X : 0 = x(0) \leq x(t) \leq x(1) = 1 \text{ for all } 0 \leq t \leq 1\}.$$

The mapping $T : C \to C$ defined by $(Tx)(t) = tx(t)$ for $0 \leq t \leq 1$ is not only nonexpansive, but even contractive. Nevertheless, it is fixed point free.

Even isometries do not necessarily have fixed points.

Example 1.6. Let $X = c_0$ and $C = \{x \in c_0 : 0 \leq x_n \leq 1\}$. The mapping $T$ defined by

$$T(x_1,x_2,\ldots) = (1,x_1,x_2,\ldots)$$

is a fixed point free isometry of $C$ into itself.

Even if a nonexpansive mapping $T$ has a unique fixed point, the sequence of iterates $\{T^n x\}$ need not converge to it.

Example 1.7. Let $C$ be the unit ball of $\ell^2$, and let the sequence of real numbers $\{a_n\}$ satisfy $0 \leq a_n \leq 1$ and $\prod_{n=1}^{\infty} a_n > 0$. Consider the linear mapping $T : C \to C$ defined by

$$T(x_1,x_2,\ldots) = (0,a_1x_1,a_2x_2,\ldots).$$

Its only fixed point is the origin, but the sequence of iterates $\{T^n e\}$ with $e = (1,0,0,\ldots)$ does not converge strongly to 0. (It does converge weakly.)

Summing up, we may say that even such "nice" spaces as bounded closed convex subsets of Banach spaces are not regular enough to provide us with very general fixed point theorems for nonexpansive or contractive mappings. As we shall see later, positive results can be established if such sets possess some additional geometric properties.

## 2. UNIFORM CONVEXITY

The purpose of this section is to discuss the concept of uniform convexity in Banach spaces. This concept was introduced by J.A. Clarkson [26] in 1936 and has since turned out to be very useful in Banach space and operator theory.

The idea behind it is an attempt to classify Banach spaces according to the degree of rotundity of their unit balls.

Recall that a Banach space $(X, |\cdot|)$ is called strictly convex (or rotund) if its unit sphere $S = \{x \in X : |x| = 1\}$ does not contain any linear segment. In other words, X is strictly convex if the following implication holds:

$$\left.\begin{array}{l} |x| = 1 \\ |y| = 1 \\ |(x+y)/2| = 1 \end{array}\right\} \Rightarrow x = y .$$

For example, the plane $R^2$ is strictly convex when equipped with the Euclidean norm $|(a,b)| = (a^2+b^2)^{1/2}$, but it is not strictly convex if it has the norms

$$|(a,b)| = |a| + |b| \text{ or } |(a,b)| = \max(|a|, |b|).$$

In order to measure the degree of strict convexity (or rotundity) of X, we define its modulus of convexity $\delta : [0,2] \to [0,1]$ by

$$\delta(\varepsilon) = \inf\{1 - |x+y|/2 : |x| \leqslant 1, |y| \leqslant 1, \text{ and } |x-y| \geqslant \varepsilon\}.$$

Roughly speaking, $\delta$ measures how deeply the midpoint of the linear segment joining two points in the unit ball B of X must lie within B.

This function is increasing on [0,2], and continuous on [0,2) (but not necessarily at $\varepsilon = 2$). It is also clear that $\delta(0) = 0$, and that $\delta(2) = 1$ if and only if X is strictly convex.

Define the characteristic of convexity $\varepsilon_0$ of X by

$$\varepsilon_0 = \sup\{0 \leq \varepsilon \leq 2 : \delta(\varepsilon) = 0\}.$$

The modulus of convexity $\delta$ is in fact strictly increasing on $[\varepsilon_0, 2]$.

The definition of $\delta$ shows that

$$\left.\begin{array}{l} |x| \leq 1 \\ |y| \leq 1 \\ |x-y| \geq \varepsilon \end{array}\right\} \Rightarrow \qquad |(x+y)/2| \leq 1-\delta(\varepsilon).$$

We shall often use this implication in the following more general form:

$$\left.\begin{array}{l} |a-x| \leq r \\ |a-y| \leq r \\ |x-y| \geq \varepsilon r \end{array}\right\} \Rightarrow \qquad |a - (x+y)/2| \leq (1-\delta(\varepsilon))r.$$

For example, if $X = R^2$ is furnished with one of the norms $|(a,b)| = |a| + |b|$ or $|(a,b)| = \max(|a|, |b|)$, then $\delta(\varepsilon) = 0$ for all $0 \leq \varepsilon \leq 2$ and $\varepsilon_0 = 2$.

If X is a Hilbert space H, then the parallelogram law leads to

$$\delta(\varepsilon) = \delta_H(\varepsilon) = 1 - (1 - \varepsilon^2/4)^{1/2} .$$

In this case $\delta(\varepsilon) > 0$ for all positive $\varepsilon$ and $\varepsilon_0 = 0$.

If $R^2$ is equipped with the norm

$$|(a,b)| = \max\{|b|, |a + b/\sqrt{3}|, |a - b/\sqrt{3}|\},$$

then the unit ball is a regular hexagon, $\varepsilon_0 = 1$, and

$$\lim_{\varepsilon \to 2} \delta(\varepsilon) = \delta(2) = 1/2 .$$

If we renorm the sequence space $c_0$ by

$$\|x\| = |x|_\infty + \left(\sum_{n=1}^{\infty} (x_n/2^n)^2\right)^{1/2} ,$$

where $|x|_\infty$ is the usual max norm, then the space becomes strictly convex and $\delta(2) = 1$, but we still have

$$\delta(\varepsilon) = 0 \text{ for all } 0 \leq \varepsilon < 2.$$

A Banach space X is said to be uniformly convex (or uniformly rotund) if $\delta(\varepsilon) > 0$ for all positive $\varepsilon$. In other words, X is uniformly convex if its characteristic of convexity $\varepsilon_0 = 0$.

Roughly speaking, this means that if two points in the unit ball of a uniformly convex space are far apart, then their midpoint must be well within it.

Any uniformly convex space is strictly convex, and a finite dimensional strictly convex space is, in fact, uniformly convex. On the other hand, there are many infinite dimensional Banach spaces which are strictly convex but not uniformly convex.

The simplest example of a uniformly convex space is Hilbert space. In fact, it is known (Nördlander [70]) that Hilbert space is the "most" uniformly convex of all Banach spaces in the sense that for any Banach space with dimension $\geq 2$,

$$\delta(\varepsilon) \leq \delta_H(\varepsilon) = 1 - (1 - \varepsilon^2/4)^{1/2} \quad \text{for all } 0 \leq \varepsilon \leq 2.$$

(For the real line, $\delta(\varepsilon) = \varepsilon/2$.)

Other examples of uniformly convex spaces include the sequence spaces $\ell^p$, the Lebesgue spaces $L^p(\Omega)$, and the Sobolev spaces $W^{m,p}(\Omega)$, where $\Omega$ is any open subset of a real Euclidean space, $m$ is a nonnegative integer, and $1 < p < \infty$.

The moduli of convexity $\delta_p$ of $L^p$ were computed precisely by Hanner [47]. If $2 \leq p < \infty$, then

$$\delta_p(\varepsilon) = 1 - (1 - (\varepsilon/2)^p)^{1/p} = (1/p)(\varepsilon/2)^p + o(\varepsilon^p)$$

for small $\varepsilon > 0$; if $1 < p \leq 2$, then

$$\delta_p(\varepsilon) = (p-1)\varepsilon^2/8 + o(\varepsilon^2) .$$

If a Banach space is renormed (by an equivalent norm), then both its modulus of convexity $\delta$ and its characteristic of convexity $\varepsilon_0$ change. This fact is already brought out by the two-dimensional examples mentioned above.

In this connection it is of interest to quote a remarkable theorem of P. Enflo [33]. Recall that a Banach space for which $\varepsilon_0 < 2$ is said to be uniformly non-square. Enflo has shown that if a Banach space is isomorphic to a uniformly non-square space, then it is also isomorphic to a uniformly convex space.

According to a well-known result of D.P. Milman [68] and B.J. Pettis [73], every uniformly convex Banach space is reflexive. There are, however, reflexive spaces which are not even isomorphic to uniformly convex spaces (Day [27]; a space is isomorphic to a uniformly convex space if and only it is super-reflexive.) Although the following result is of course valid in all reflexive Banach spaces, we shall present a direct proof for uniformly convex spaces. We shall later use an analog of this proof in our study of the hyperbolic metric on the Hilbert ball. Let

$$d(z,C) = \inf\{|z-x| : x \in C\}$$

denote the distance from a point $z \in X$ to a subset $C$ of a Banach space $X$.

Theorem 2.1. Let $\{C_n : n = 1,2,...\}$ be a decreasing sequence of nonempty bounded closed convex subsets of a uniformly convex Banach space $X$. Then the intersection $\cap\{C_n : n = 1,2,...\}$ is a nonempty closed convex subset of $X$.

Proof. Let $z$ be a point in $X$ which does not belong to $C_1$, $r_n = d(z,C_n)$ and $r = \lim_{n\to\infty} r_n$. Also, let $\{p_n\}$ be a positive sequence that decreases to zero, $D_n = \{x \in C_n : |z-x| \le r + p_n\}$, and $d_n$ the diameter of $D_n$. If $x$ and $y$ belong to $D_n$ and $|x-y| \ge d_n - p_n$, then

$$|z - (x + y)/2| \le (1 - \delta(|x-y|/(r + p_n)))(r + p_n),$$

and

$$r_n \le (1 - \delta((d_n - p_n)/(r + p_n)))(r + p_n).$$

Denoting $\lim_{n\to\infty} d_n$ by $d$, we obtain a contradiction unless $d = 0$. This in turn implies that $\cap\{D_n : n = 1,2,...\}$ is nonempty, and so is $\cap\{C_n : n = 1,2,...\}$.

This intersection theorem remains valid if the sequence $\{C_n\}$ is replaced by an arbitrary decreasing net of nonempty bounded closed convex sets. It is not valid, however, in all Banach spaces. To see this, let $X = C[0,1]$, for example, and

$$C_n = \{x \in C[0,1] : 0 \le x(t) \le t^n \text{ for all}$$

$$0 \le t \le 1, \text{ and } x(1) = 1\}.$$

Let C be a convex subset of a Banach space. Recall that a function $f : C \to R$ is said to be convex if

$$f((1-t)x + ty) \leqslant (1-t)f(x) + tf(y)$$

for all x and y in C and all $0 \leqslant t \leqslant 1$. Our next result is a direct consequence of Theorem 2.1.

Proposition 2.2. Let C be a closed convex subset of a uniformly convex Banach space X, and let $f : C \to [0,\infty)$ be a convex function. If f is lower semicontinuous and

$$f(x) \to \infty \text{ as } |x| \to \infty,$$

then f attains its minimum on C. If, in addition, $f((x+y)/2) < \max\{f(x), f(y)\}$ for all $x \neq y$, then f attains its minimum at exactly one point.

Proof. Denote the infimum of f on C by I, and apply Theorem 2.1 to the sets $C_n = \{x \in C : f(x) \leqslant I + 1/n\}$.

A function $f : C \to R$ is said to be quasi-convex if

$$f((1-t)x + ty) \leqslant \max\{f(x), f(y)\}$$

for all x and y in C and $0 \leqslant t \leqslant 1$. Proposition 2.2 is valid for quasi-convex functions too.

## 3. NEAREST POINT PROJECTIONS

Recall that a subset D of a Banach space X is said to be a Chebyshev set if to each point x in X there corresponds a unique point z in D such that $|x-z| = d(x,D)$. In this case we may define the nearest point projection $P : X \to D$ by assigning z to x. The following result is an application of Proposition 2.2.

Proposition 3.1. Every closed convex subset of a uniformly convex Banach space is a Chebyshev set.

Proof. Let C be a closed convex subset of a uniformly convex Banach space X, and apply Proposition 2.2 to the function $f : C \to [0,\infty)$ defined by $f(y) = |x-y|$.

Proposition 3.2. The nearest point projection onto a closed convex subset of a uniformly convex Banach space is continuous.

Proof. Let C be a closed convex subset of a uniformly convex Banach space X, and let $P : X \to C$ be the nearest point projection onto C. Let $x_n$ converge to x in X and denote $Px_n$ by $z_n$. We may assume that x does not belong to C. If $\{z_n\}$ is not a Cauchy sequence, then there are a positive $\varepsilon$ and subsequences $\{z_{m_k}\}$ and $\{z_{n_k}\}$ such that $m_k < n_k$ and $|z_{m_k} - z_{n_k}| \geq \varepsilon$ for all k. Denote $z_{m_k}$ and $z_{n_k}$ by $u_k$ and $v_k$ respectively, and let $M_k = \max\{|x-u_k|, |x-v_k|\}$. Note that $M_k \to d(x,C)$ as $k \to \infty$. We have

$$d(x,C) \leq |x - (u_k + v_k)/2| \leq M_k(1 - \delta(|u_k - v_k|/M_k)),$$

and

$$\delta(\varepsilon/M_k) \leq 1 - d(x,C)/M_k .$$

Letting $k \to \infty$, we see that $\varepsilon$ cannot be positive. Thus $\{z_n\}$ is a Cauchy sequence and therefore converges to a point z in C. Since $|x-z| = d(x,C)$, $z = Px$ and the proof is complete.

Although Proposition 3.1 is valid in any strictly convex reflexive space, Proposition 3.2 is not, even if C is a subspace. Examples of such discontinuous nearest point projections have been constructed by Kripke [61] and Brown [16].

Recall that the (normalized) duality map from a (real) Banach space X into the family of nonempty (by the Hahn-Banach theorem) weak-star compact convex subsets of its dual $X^*$ is defined by

$$J(x) = \{x^* \in E^* : (x,x^*) = |x|^2 \text{ and } |x^*| = |x|\}$$

for each x in X.

In order to proceed, we shall need the following simple fact concerning the convex function $f: [0,1] \to [0,\infty)$ defined by $f(s) = |u + sv|$, where u and v are two points in X.

Lemma 3.3. For two points u and v in a real Banach space X, the following are equivalent:

(a) $|u| \leq |u + sv|$ for all $0 \leq s \leq 1$;

(b) $|u + sv|$ increases on $[0,1]$;

(c) there is $j \in J(u)$ such that $(v,j) \geq 0$.

We use this lemma in order to characterize the nearest point projection onto a convex Chebyshev set.

Proposition 3.4. Let C be a convex Chebyshev set in a real Banach space X, x a point in X, z a point in C, and $P : X \to C$ the nearest point projection of X onto C.

Then the following are equivalent:

(a) $z = Px$;

(b) for each y in C,

$|x-z| = \min \{|x - (1-t)y - tz| : 0 \leq t \leq 1\}$;

(c) for each y in C,

$|x - (1-t)y - tz|$ decreases on $[0,1]$;

(d) for each y in C, there is

$j \in J(x-z)$ such that $(y-z,j) \leq 0$.

Proof. It is clear that (a) and (b) are equivalent. Choosing $u = x-z$ and $v = z-y$, we can use Lemma 3.3 to show that (b), (c) and (d) are also equivalent.

The norm of X is said to be Gâteaux differentiable (and X is said to be smooth) if

$$\lim_{t\to 0} (|x+ty| - |x|)/t \tag{3.1}$$

exists for each x and y in the unit sphere $U = \{x \in X: |x| = 1\}$ of X. In this case the duality map J is single-valued and the limit (3.1) is equal to (y,Jx). Proposition 3.4 shows that if X is smooth, then the nearest point projection of X onto a convex Chebyshev set C is characterized by

$$(y - Px, J(x - Px)) \leq 0, \tag{3.2}$$

where $x \in X$ and $y \in C$.

Note that if $X^*$ is strictly convex (smooth), then X is smooth (strictly convex). Hence, if X is reflexive, X is smooth (strictly convex) if and only if $X^*$ is strictly convex (smooth). However this duality is not valid for all Banach spaces.

We have been assuming that X is real only for simplicity. If X is complex, then (3.2), for example, should be replaced by

$$\mathrm{Re}(y - Px, J(x-Px)) \leq 0. \tag{3.3}$$

This remark also applies to the following discussion of the Hilbert space case.

In the special, but important, case of a Hilbert space, the duality map is simply the identity operator, and we obtain the following result.

Proposition 3.5. Let C be a closed convex subset of a (real) Hilbert space H, x a point in H, z a point in C, and $P : H \to C$ the nearest point projection of H onto C. Then the following are equivalent:

(a) $z = Px$;

(b) for each y in C,

$$|x-z| = \min\{|x - (1-t)y - tz| : 0 \leq t \leq 1\};$$

(c) for each y in C,

$$|x - (1-t)y - tz| \text{ decreases on } [0,1];$$

(d) $(y-z, x-z) \leq 0$ for all y in C;

(e) for each y in C,

$$|y-z| = \min\{|y-(1-t)x-tz| : 0 \leq t \leq 1\};$$

(f) for each y in C,

$$|y-(1-t)x-tz| \text{ decreases on } [0,1];$$

(g) $|y - (2z-x)| \leq |y-x|$ for all y in C.

Proof. We already know that (a), (b), (c) and (d) are all equivalent. In any smooth Banach space, (e) and (f) are equivalent to the requirement that $(x-z, J(y-z)) \leq 0$ for all y in C. But in Hilbert space this condition coincides with (d). Finally, note that in Hilbert space, the inequality $|w - v| \leq |w - u|$ is equivalent to $(w - (v+u)/2, v-u) \geq 0$. Hence (g) is equivalent to (d).

Let u and v be two points in a Hilbert space H. As we have seen, the equidistant set

$$Eq(u,v) = \{w \in H : |w-v| = |w-u|\}$$

is the hyperplane

$$\{w \in H: (w-(v+u)/2, v-u) = 0\},$$

and the half-space

$$\overset{+}{Eq}(u,v) = \{w \in H: |w-v| \leq |w-u|\}$$

$$= \{w \in H: (w - (v+u)/2, v-u) \geq 0\}.$$

Part (g) of Proposition 3.5 shows that Px is the unique point z in C such that C is contained in the half-space $E_q^+(x,2z-x)$ determined by x and by 2z−x. The point 2Px − x may be thought of as the "reflection" of x in C.

We continue with an important consequence of Proposition 3.5.

<u>Theorem 3.6.</u> The nearest point projection onto a closed convex subset of a Hilbert space is nonexpansive.

<u>Proof.</u> Let C be a closed convex subset of a Hilbert space H. The nearest point projection $P : H \to C$ exists by Proposition 3.1. By part (d) of Proposition 3.5,

$$(y - Px, x - Px) \leqslant 0 \tag{3.4}$$

for all x in H and y in C. Now let z and w be two arbitrary points in H. Since $(Pw - Pz, z - Pz) \leqslant 0$ and $(Pz - Pw, w - Pw) \leqslant 0$, it follows that $(Pz - Pw, Pz - Pw + w-z) \leqslant 0$. Therefore $|Pz - Pw|^2 \leqslant |Pz - Pw||z - w|$ and the proof is complete.

Alternatively, Theorem 3.6 can be established by using the nearest point projection Q onto the line determined by Pz and Pw. It is easy to see that the nearest point projection onto any line is nonexpansive. It is also not difficult to see that the segment with Pz and Pw as endpoints is contained in the segment determined by Qz and Qw. Hence $|Pz - Pw| \leqslant |Qz - Qw| \leqslant |z-w|$.

Let $P : H \to C$ be the nearest point projection of a Hilbert space onto a closed convex subset C of H. If C is a subspace, then P is the familiar orthogonal projection. If C is the closed unit ball, then P is the radial projection R defined by

$$Rx = \begin{cases} x & \text{if} \quad |x| \leqslant 1 \\ x/|x| & \text{if} \quad |x| \geqslant 1 . \end{cases}$$

This radial projection is the nearest point projection onto the closed unit ball of any strictly convex space, but it is no longer nonexpansive. In fact, if the dimension of the space is not less than 3, then the radial projection is never nonexpansive outside Hilbert space [34]. We shall discuss the existence of nonexpansive retractions later.

We conclude the present section with several additional properties of the nearest point projection in Hilbert space.

From the proof of Theorem 3.6 we see that

$$(Px - Py,\ x-y) \geq |Px - Py|^2 \tag{3.5}$$

for all x and y in H. Hence

$$(Px - Py,\ x - y) \geq 0. \tag{3.6}$$

We recall that mappings which satisfy (3.6) are said to be monotone. Let I denote the identity operator. It is not difficult to see that (3.5) implies (and in fact is equivalent to) the nonexpansiveness of $S = 2P - I$. Hence $I - P = (-S+I)/2$ is nonexpansive too.

Parts (d) of Propositions 3.4 and 3.5 show that each point on the ray $\{Px + t(x - Px) : t \geq 0\}$ is mapped by P back onto Px. In other words,

$$P(Px + t(x - Px)) = Px \tag{3.7}$$

for all $t \geq 0$. Retractions that satisfy (3.7) are said to be sunny. Hence

$$|Px - Py| = |P((1-t)x + tPx) - P((1-t)y + tPy)|$$

$$\leq |(1-t)x + tPx - ((1-t)y + tPy)| \text{ for all } 0 \leq t \leq 1,$$

and the convex function

$$|(1-t)x + tPx - ((1-t)y + tPy)|$$

decreases on [0,1]. Mappings with this property are called firmly nonexpansive.

This class of nonexpansive mappings will be discussed later.

Summing up, we see that the nearest point mapping in Hilbert space is sunny, firmly nonexpansive, and monotone.

## 4. ASYMPTOTIC CENTERS

Let $\{x_n : n = 1,2,\ldots\}$ be a bounded sequence in a Banach space X, and let C be a closed convex subset of X. Consider the functional $f : X \to [0,\infty)$ defined by $f(x) = \limsup_{n\to\infty} |x_n - x|$ for x in X. In order to emphasize the dependence of f on the sequence $\{x_n\}$, we shall denote this functional by $r(x,\{x_n\})$. The infimum of f(x) over C is called the asymptotic radius of $\{x_n\}$ with respect to C and is denoted by $r(C, \{x_n\})$. A point z in C is said to be an asymptotic center of the sequence $\{x_n\}$ with respect to C if $f(z) = \min \{f(x) : x \in C\}$. The set of all asymptotic centers is denoted by $A(C, \{x_n\})$. Thus we have

$$r(C, \{x_n\}) = \inf\{r(x,\{x_n\}) : x \in C\}$$

and

$$A(C,\{x_n\}) = \{z \in C : r(z,\{x_n\}) = r(C,\{x_n\})\}.$$

Note that these concepts can be defined for any metric space M and any subset of M.

If the sequence $\{x_n\}$ converges to x in C, then

$$r(C,\{x_n\}) = 0$$

and

$$A(C, \{x_n\}) = \{x\}.$$

If $\{x_n\}$ converges to a point x which is not in C, then

$$r(C,\{x_n\}) = d(x,C)$$

and

$$A(C, \{x_n\}) = \{z \in C : |z-x| = d(x,C)\}.$$

This set may be empty, a singleton, or contain infinitely many points. If $\{x_n\}$ is a periodic sequence with only two values u,v in C, then

$$r(C, \{x_n\}) = |u-v|/2$$

and

$$A(C,\{x_n\}) = \{z \in C : |u-z| = |v-z| = |u-v|/2\}.$$

The point $(x+y)/2$ always belongs to this set, but $A(C, \{x_n\})$ may contain other points if X is not strictly convex. The situation is simpler in uniformly convex spaces.

Theorem 4.1. Every bounded sequence in a uniformly convex Banach space X has a unique asymptotic center with respect to any closed convex subset of X.

Proof. Let $\{x_n\}$ be a bounded sequence in X, C a closed convex subset of X, and $f : C \to [0,\infty)$ the functional defined by $f(x) = \limsup_{n\to\infty} |x_n - x|$. This continuous functional is convex, and $f(x) \to \infty$ as $|x| \to \infty$. Therefore, in order to apply Proposition 2.2 all we have to show is that

$$f((x+y)/2) < \max\{f(x),f(y)\} \text{ for all } x \neq y.$$

To this end, let $M = \max\{f(x), f(y)\}$, and note that for each positive $\varepsilon$, there is an $N(\varepsilon)$ such that $|x_n - x| \leq f(x) + \varepsilon$ and

$$|x_n - y| \leq f(y) + \varepsilon \text{ for all } n \geq N(\varepsilon).$$

Therefore

$$|x_n - (x+y)/2| \leq (M + \varepsilon)(1 - \delta(|x-y|/(M + \varepsilon)))$$

and

$$f((x+y)/2) \leq M(1 - \delta(|x-y|/M)) < M.$$

This completes the proof.

We continue with two important properties of the asymptotic center in Hilbert space.

Theorem 4.2. In a Hilbert space H, the weak limit of a weakly convergent sequence coincides with its asymptotic center with repect to H.

Proof. Let the sequence $\{x_n\}$ converge weakly to $z \in H$. If y is any other point in H,

$$|x_n - y|^2 = |x_n - z|^2 + 2\text{Re}(x_n - z, z-y) + |z-y|^2.$$

Hence

$$\limsup_{n\to\infty} |x_n - y|^2 = \limsup_{n\to\infty} |x_n - z|^2 + |z-y|^2,$$

and the result follows.

In the setting of Theorem 4.2, it is clear that the asymptotic center of $\{x_n\}$ belongs to the closed convex hull of $\{x_n\}$. As a matter of fact, this is always true in Hilbert space. We shall establish this claim by appealing to Theorem 3.6.

Theorem 4.3. In a Hilbert space H, the asymptotic center with respect to H of a bounded sequence $\{x_n\}$ belongs to the closed convex hull of $\{x_n\}$.

Proof. Let z be the asymptotic center of $\{x_n\}$ with respect to H, and let P denote the nearest point projection from H onto the closed convex hull of $\{x_n\}$. Since

$$|x_n - Pz| = |Px_n - Pz| \leq |x_n - z|,$$

we see that Pz must coincide with z.

Both Theorem 4.2 and Theorem 4.3 are not true in all uniformly convex spaces. Although Theorem 4.2 is valid in all the sequence spaces $\ell^p$, $1 < p < \infty$, it does not hold in the Lebesgue spaces $L^p$, $p \neq 2$ [72]. Theorem 4.3 does hold in all two-dimensional strictly convex spaces, but it is not valid even in the three-dimensional space $\ell_3^p$, $p \neq 2$, as the following example shows:

Let $X = \ell_3^p$, $x_1 = (1,0,0)$, $x_2 = (0,1,0)$, and $x_3 = (0,0,1)$. The asymptotic center of the periodic sequence with the values $\{x_1,x_2,x_3\}$ is $t(1,1,1)$ with $t = (2^{1/p-1} + 1)^{-1}$. Therefore it belongs to the closed convex hull of $\{x_1,x_2,x_3\}$ if and only if $t = 1/3$ and $p = 2$.

Remark. As a matter of fact, we can extend Theorem 4.2. let C be a closed convex subset of a Hilbert space H, $P: H \to C$ the nearest point projection onto C, and $\{x_n\} \subset H$ a sequence that converges weakly to x. Then the asymptotic center of $\{x_n\}$ with respect to C is Px. Indeed let y belong to C. We have
$|x_n - y|^2 = |x_n - Px|^2 + 2\mathrm{Re}(x_n - Px, Px - y) + |Px - y|^2 =$

$|x_n - Px|^2 + 2\mathrm{Re}(x_n - x, Px - y) + 2\mathrm{Re}(x - Px, Px - y) + |Px - y|^2$
$\geq |x_n - Px|^2 + 2\mathrm{Re}(x_n - x, Px - y) + |Px - y|^2$. Hence

$$\limsup_{n\to\infty} |x_n - y|^2 = \limsup_{n\to\infty} |x_n - Px|^2 + |Px - y|^2,$$

and the result follows.

In the following sections we shall use the concept of an asymptotic center (due to M. Edelstein) to study nonexpansive mappings and their fixed points.

## 5. FIXED POINTS IN UNIFORMLY CONVEX SPACES

Let C be a bounded closed convex subset of a Banach space X, and let $T : C \to C$ be a nonexpansive mapping. As we have already observed in Section 1, the fixed point set $F(T) = \{x \in C: Tx = x\}$ may be empty. The quest for geometric conditions on C which will guarantee the existence of at least one fixed point for each nonexpansive self-mapping of C has led to an extensive fixed point theory for nonexpansive mappings. We shall discuss several facets of this theory in this and in the next section.

The basic result in uniformly convex spaces is independently due to Browder [13], Göhde [44], and Kirk [56].

<u>Theorem 5.1.</u> Let C be a bounded closed convex subset of a uniformly convex Banach space X. If $T : C \to C$ is nonexpansive, then it has a fixed point.

We shall present three proofs of this theorem. The first is based on the intersection method; the second and the third use asymptotic centers.

<u>Proof 1.</u> For a positive $\varepsilon$, let $C_\varepsilon = \{x \in C : |x - Tx| \leq \varepsilon\}$. Each $C_\varepsilon$ is closed and nonempty by Proposition 1.4. If x and y belong to $C_\varepsilon$ and $z = (x+y)/2$, then

$$|Tz - x| \leq |Tz - Tx| + \varepsilon \leq |x - y|/2 + \varepsilon = R_\varepsilon$$

and $|Tz - y| \leq R_\varepsilon$.

Hence

$$|z - Tz| \leq (1 - \delta(|x - y|/R_\varepsilon))R_\varepsilon ,$$

where $\delta$ is the modulus of convexity of X. If

$$|x-y| \leq \sqrt{\varepsilon} \text{ , then } |z - Tz| \leq \sqrt{\varepsilon}/2 + \varepsilon.$$

Otherwise $|z - Tz| \leq \left(1 - \delta\left(\frac{2}{1+2\sqrt{\varepsilon}}\right)\right)(d/2 + \varepsilon)$, where d is the diameter of C. Therefore we may conclude that $z \in C_{p(\varepsilon)}$, where $\lim_{\varepsilon \to 0} p(\varepsilon) = 0$. Now let $u \in C$, $r_\varepsilon = d(u, C_\varepsilon)$, and $r = \lim_{\varepsilon \to 0} r_\varepsilon$. Also, let $s(\varepsilon)$ be a positive function that decreases to 0 as $\varepsilon \to 0$,

$$D_\varepsilon = \{x \in C_\varepsilon : |u - x| \leq r + s(\varepsilon)\},$$

and $d_\varepsilon$ the diameter of $D_\varepsilon$. If x and y are two points in $D_\varepsilon$ such that $|x-y| \geq d_\varepsilon - s(\varepsilon)$, then

$$r_{p(\varepsilon)} \leq |u-(x+y)/2| \leq \left(1-\delta\left(\frac{d_\varepsilon - s(\varepsilon)}{r + s(\varepsilon)}\right)\right)(r+s(\varepsilon)).$$

Letting $\varepsilon \to 0$, we obtain a contradiction unless $\lim_{\varepsilon \to 0} d_\varepsilon = 0$. This implies that $\cap \{D_\varepsilon : \varepsilon > 0\}$ is nonempty, and so is

$$F(T) = \cap \{C_\varepsilon : \varepsilon > 0\}.$$

Proof 2. Let $x \in C$ and let z be the asymptotic center of $\{T^n x\}$ with respect to C. We have $|T^n x - Tz| \leq |T^{n-1} x - z|$ for all $n \geq 1$. Since the asymptotic center is unique, Tz must coincide with z.

Proof 3. Let $\{y_n\} \subset C$ satisfy $y_n - Ty_n \to 0$. If $z$ is the asymptotic center of $\{y_n\}$ with respect to $C$, then

$$|y_n - Tz| \leq |y_n - Ty_n| + |Ty_n - Tz| \leq |y_n - Ty_n| + |y_n - z|.$$

Hence

$$\limsup_{n\to\infty} |y_n - Tz| \leq \limsup_{n\to\infty} |y_n - z|$$

and $z = Tz$.

What happens when $C$ is not bounded? One answer is provided by our next result, which follows from Proofs 2 and 3 of Theorem 5.1.

Theorem 5.2. Let $C$ be a closed convex subset of a uniformly convex Banach space. If $T : C \to C$ is nonexpansive, then the following are equivalent:

(a) $T$ has a fixed point;

(b) There is a point $x$ in $C$ such that the sequence of iterates $\{T^n x\}$ is bounded;

(c) There is a bounded sequence $\{y_n\} \subset C$ such that $\lim_{n\to\infty} (y_n - Ty_n) = 0$.

Theorems 5.1 and 5.2 are no longer true if the Banach space $X$ is merely strictly convex. To see this, consider the space $X = C[0,1]$ renormed by the norm

$$\|x\| = \max\{|x(t)| : 0 \leq t \leq 1\} + \left(\int_0^1 x^2(t)dt\right)^{1/2}.$$

Once again, let

$$C = \{x \in X : 0 = x(0) \leq x(t) \leq x(1) = 1 \text{ for all } 0 \leq t \leq 1\}.$$

The space $X$ is strictly convex, but the nonexpansive mapping $T : C \to C$ defined by $(Tx)(t) = tx(t)$ is fixed point free.

However, the following fact is true in all strictly convex spaces.

Proposition 5.3. Let C be a closed convex subset of a strictly convex Banach space. If $T : C \to C$ is nonexpansive, then the fixed point set F(T) of T is closed and convex.

Proof. F(T) is closed because T is continuous. If $x = Tx$, $y = Ty$, and $z = (x+y)/2$, then $|Tz - x| = |Tz - Tx| \leq |z - x| = |x - y|/2$ and $|Tz - y| \leq |x - y|/2$. Strict convexity now implies that $z = Tz$.

Let X be a uniformly convex Banach space, C a closed convex subset of X, and $T : C \to C$ a nonexpansive mapping with a nonempty fixed point set F(T). It is easy to see that the asymptotic center of a bounded sequence in F(T) must again be a fixed point of T. In other words, the closed convex set F(T) is closed with respect to asymptotic centers. It is of interest to observe that not all closed convex subsets of C have this property. For example, let $C = X = \ell_3^p$. We have seen that for $p \neq 2$ the asymptotic center with respect to C of a bounded sequence in the plane

$$F = \{(x_1, x_2, x_3) \in X : x_1 + x_2 + x_3 = 1\}$$

need not belong to F. Therefore this closed convex set is not the fixed point set of a nonexpansive mapping $T : X \to X$. Other properties of the fixed point sets of nonexpansive mappings will be discussed later.

## 6. THE FIXED POINT PROPERTY FOR NONEXPANSIVE MAPPINGS

We shall say that a closed convex subset C of a Banach space X has the fixed point property for nonexpansive mappings (FPP for short) if every nonexpansive $T: C \to C$ has a fixed point. We have just seen that if X is uniformly convex and C is bounded, then C has the

FPP. In this section we intend to discuss the FPP in other Banach spaces.

Let $T : C \to C$ be nonexpansive, and let D be a nonempty closed convex subset of C which is invariant under T. We shall say that D is a minimal T-invariant set (or simply minimal) if it contains no proper closed convex subsets which are invariant under T.

Lemma 6.1. Let T be a nonexpansive self-mapping of a closed convex subset C of a Banach space X. If C is weakly compact, then it contains a minimal T-invariant set.

Proof. The family of all nonempty closed convex subsets of C which are invariant under T can be partially ordered by inclusion. Since C is weakly compact, the finite intersection property can be used to show that this family satisfies the hypotheses of Zorn's Lemma. The result follows.

We now list several properties of the minimal sets obtained in Lemma 6.1. Let clco(D) denote the closed convex hull of a set D.

Property 1. If C is minimal, then $\mathrm{clco}(T(C)) = C$.

Proof. The closed convex set $\mathrm{clco}(T(C))$ is invariant under T.

A point x in a set C with positive diameter d is said to be diametral if $\sup\{|x-y| : y \in C\} = d$. There are closed convex sets which consist entirely of diametral points.

Example 6.2. Let $C=\{x \in C[0,1]: 0=x(0) \leq x(t) \leq x(1) = 1\}$. In this case $\mathrm{diam}(C) = 1$ and $\lim\limits_{n\to\infty} |x - t^n| = 1$ for all $x \in C$.

Example 6.3. For $r > 1$, let $X_r$ be the space $\ell^2$ renormed by

$$|x|_r = \max\{\|x\|_2, r\|x\|_\infty\} ,$$

where $\|\cdot\|_2$ denotes the $\ell^2$ norm and $\|\cdot\|_\infty$ denotes the $\ell^\infty$ norm. Consider the set $C = \{x = (x_1, x_2, \ldots) \in X_r : x_j \geq 0$ for all $j$ and $\|x\|_2 \leq 1\}$. For $r \geq \sqrt{2}$, $\text{diam}(C) = r$. Now let $e_n$ be the n-th unit vector in $X_r$ ($e_{nj} = 0$ for $j \neq n$, $e_{nn} = 1$). Since $\lim |x-e_n|_r = r$ for each $x$ in $C$, we see that all points of $C$ are diametral. Since $X_r$ is isomorphic to $\ell^2$, it is reflexive and $C$ is weakly compact.

Those sets which consist entirely of diametral points are called diametral.

Property 2. If $C$ is minimal, then it is diametral.

Proof. Suppose there is a point $x$ in $C$ such that

$$\sup\{|x-y| : y \in C\} = r < \text{diam}(C).$$

Consider the set

$$D = \{z \in C : |z-y| \leq r \text{ for all } y \in C\}.$$

This is a nonempty proper closed convex subset of $C$. To obtain a contradiction, we show now that $D$ is invariant under $T$. Indeed, if $y \in C$ and $\varepsilon > 0$, there is, by Property 1, a convex combination $w = \sum_{j=1}^{n} a_j Tu_j$ with $u_j \in C$ such that $|y-w| < \varepsilon$. Therefore if $z \in D$ then

$$\begin{aligned} |Tz-y| &\leq |Tz-w| + |w-y| \\ &\leq \sum_{j=1}^{n} a_j |Tz-Tu_j| + \varepsilon \\ &\leq \sum_{j=1}^{n} a_j |z-u_j| + \varepsilon \leq r+\varepsilon. \end{aligned}$$

Since this inequality is true for all $\varepsilon > 0$, $Tz$ does belong to $D$.

Property 3. If C is minimal, then $\limsup_{n\to\infty}|T^n x-y| = \text{diam}(C)$ for all x and y in C.

Proof. Let x belong to C. Since C is minimal,

$$\limsup_{n\to\infty} |T^n x-y|$$

is the same for all y in C. If this limit $r < \text{diam}(C)$, consider the intersections of C with the closed balls centered at the points of C with radii $r < R < \text{diam}(C)$. This family of weakly closed sets has the finite intersection property. Therefore the intersection of all members of the family is nonempty. But any point in this intersection is not diametral. This contradiction shows that $r = \text{diam}(C)$.

Property 4. Let $\{y_n\} \subset C$ satisfy $y_n - Ty_n \to 0$. If C is minimal, then $\lim_{n\to\infty} |y_n - x| = \text{diam}(C)$ for all x in C.

Proof. The proof of this fact is similar to the proof of Property 3. We can conclude that $\lim_{n\to\infty} |y_n - x| = \text{diam}(C)$ because every subsequence $\{z_n\}$ of $\{y_n\}$ also satisfies $z_n - Tz_n \to 0$.

In the notation of Section 4, Properties 3 and 4 may be expressed in the following way:

$$r(C, \{T^n x\}) = \text{diam}(C) = r(C, \{y_n\}),$$

$$A(C, \{T^n x\}) = C = A(C, \{y_n\}).$$

Convex diametral sets are in some sense pathological. Therefore it is natural to consider the following concept [12]: A closed convex set C is said to have normal structure if it does not contain bounded closed convex diametral subsets of positive diameter. In view of Property 2, we immediately obtain the following fixed point theorem due to W.A. Kirk [56].

Theorem 6.4. If a weakly compact convex subset C of a Banach space has normal structure, then it has the FPP.

As we shall see in the next section, all uniformly convex spaces have normal structure. All compact C also have this property. Thus normal structure of the set C is a weaker assumption than uniform convexity of the underlying space.

On the other hand, as we have seen in Example 6.3, not all weakly compact convex subsets of Banach spaces possess normal structure. The question whether every weakly compact convex subset C of a Banach space X has the FPP has remained open for a long time [80]. It has been recently answered in the negative by D.E. Alspach [1] who constructed the following example. This example shows, of course, that in the setting of Lemma 6.1 there are minimal invariant sets which are not singletons.

Example 6.5. Let X be the Lebesgue space $L^1[0,1]$ and

$$C = \{x \in L^1[0,1]: \quad 0 \leq x(t) \leq 2 \quad \text{a.e. and } \int_0^1 x(t)dt = 1\}.$$

Define $T : C \to C$ by

$$(Tx)(t) = \begin{cases} \min\{2, 2x(2t)\} & 0 \leq t \leq \frac{1}{2} \\ \max\{0, 2x(2t-1)-2\} & \frac{1}{2} < t \leq 1. \end{cases}$$

C is convex and weakly compact (as a weakly closed subset of an order interval), and T is an isometry. If x is a fixed point of T, then the set

$$A = \{0 \leq t \leq 1 : x(t) = 2\} = \{0 \leq t \leq 1 : Tx(t) = 2\}$$

is of Lebesgue measure 1/2. The measure of the sets $A \cap [0,1/2]$ and $A \cap [1/2,1]$ is 1/4. In fact, it can be shown that the intersection of A with any interval has measure exactly half the measure of that interval. Hence A cannot be measurable and T is fixed point free.

Note that since $|x(t) - 1| \leq 1$ for all x in C while diam (C) = 2, C is not diametral. Hence it is not a minimal T-invariant set.

On the positive side, it has been recently shown by Maurey [67] that all weakly compact convex subsets of reflexive subspaces of $L^1[0,1]$ have the FPP. He has also proved that all the weakly compact convex subsets of the sequence space $c_0$, the Hardy space $H^1$, and the spaces $X_r$, $1 < r < \infty$, defined in Example 6.3 have the FPP. It has also been shown [7] that a condition weaker than normal structure is sufficient to guarantee the FPP for weakly compact convex sets. It is still not known, however, if every weakly compact (equivalently, bounded closed) convex subset of a reflexive space has the FPP.

What about weak-star compact convex subsets of dual spaces? Some of them have the FPP: For example, those with normal structure, subsets of $\ell^1$ [53], and balls in $L^\infty$ [91,92]. But not all of them do - the positive part of the unit ball in $\ell^1$ renormed by

$$|x| = \max\{\|x^+\|_1, \|x^-\|_1\}$$

provides us with a counterexample [63].

For more details concerning the FPP see [57,58].

We conclude this section with an example due to K. Goebel and T. Kuczumow [41]. They constructed a bounded closed convex subset C of $\ell^1$ which is not weak-star compact but has the FPP. Moreover, C contains a closed convex subset which does not possess the FPP. Thus in contrast with the properties of uniform convexity and normal structure, the FPP is not hereditary.

Example 6.6. Let $\{e_j\}$ be the standard Schauder basis in $c_0^* = \ell^1$, $\{a_j\}$ a bounded sequence of non-negative reals, and

$$f_j = (1+a_j)e_j, \; j=1,2,\ldots \, .$$

Consider the set

$$C = \{x \in \ell^1 : x = \sum_{j=1}^{\infty} \lambda_j f_j \text{ with } \lambda_j \geq 0 \text{ and } \sum_{j=1}^{\infty} \lambda_j = 1\}.$$

This set is bounded, closed, and convex, but it is not weak-star compact. In fact, the sequence $\{f_j\} \subset C$ converges weak-star to the origin which does not belong to C. The weak-star closure of C is

$$D = \{x \in \ell^1 : x = \sum_{j=1}^{\infty} \mu_j f_j \text{ with } \mu_j \geq 0 \text{ and } \sum_{j=1}^{\infty} \mu_j \leq 1\}.$$

For any x in D, let $\delta_x = 1 - \sum_{j=1}^{\infty} \mu_j$ . Also, let

$$a = \inf\{a_j : j \geq 1\}, \; N_0 = \{j \geq 1 : a_j = a\},$$

and

$$Px = \{y \in C : |x-y| = d(x,C)\}.$$

It is not difficult to see that $d(x,C) = \delta_x(1+a)$ and that Px is the closed convex hull of $\{x + \delta_x f_j : j \in N_0\}$. Now let $T : C \to C$ be nonexpansive and let $\{x_n\}$ be a sequence in C which converges weak-star to $x \in D$ and which satisfies $|x_n - Tx_n| \to 0$. Let

$$r(y) = \limsup_{n\to\infty} |x_n - y|.$$

It is not difficult to see that $r(y) = r(x) + |y-x|$. Therefore the set $A(C,\{x_n\})$ of asymptotic centers of $\{x_n\}$ with respect to C coincides with Px. This set may be empty (if $N_0$ is empty), compact (if $N_0$ is finite) or even not weak-star compact (if $N_0$ is infinite).

If $N_0$ is non-empty but finite, then $A(C,\{x_n\}) = Px$ is a compact convex subset of C which is invariant under T. Therefore it must contain a fixed point of T and C has the FPP.

It can be shown that C does not have the FPP if $N_0$ is either empty or infinite. For example, if all the $a_j$'s are equal, then the shift $(\xi_1,\xi_2,\ldots) \to (0,\xi_1,\xi_2,\ldots)$ is a fixed point free isometry of C. Consequently, if $a_1 = 0$ and all the other $a_j$'s are equal and positive, then C has the FPP, but its subset

$$\{x = (\xi_1,\xi_2,\ldots) \in C : \xi_1 = 0\}$$

does not.

## 7. STABILITY

We begin this section with the following useful result. It shows, in particular, that all uniformly convex spaces have normal structure.

<u>Theorem 7.1.</u> Let $\varepsilon_0(X)$ be the characteristic of convexity of a Banach space X. If $\varepsilon_0(X) < 1$, then X has normal structure.

<u>Proof.</u> Let C be a nonempty bounded closed convex subset of X with positive diameter d. Let $\varepsilon$ be positive. If x and y are two points in C such that $|x-y| \geq d-\varepsilon$, $z = (x+y)/2$ and $u \in C$, then $|u-z| \leq (1-\delta(\frac{d-\varepsilon}{d}))d$. Since $\varepsilon_0(X) < 1$, z will not be diametral if $\varepsilon$ is sufficiently small.

There are Banach spaces with $\varepsilon_0 = 1$ without normal structure [24]. On the other hand, there are reflexive Banach spaces with normal structure which are not super-reflexive [8]. For such spaces $\varepsilon_0 = 2$.

Theorem 6.4 now implies the following corollary.

Corollary 7.2. A bounded closed convex subset of a Banach space with $\varepsilon_0 < 1$ has the FPP.

Let $(X, |\cdot|)$ be a Banach space with $\varepsilon_0 < 1$, and let X be renormed by the equivalent norm $\|\cdot\|$. Then there are two positive constants m and M such that

$$m|x| \leqslant \|x\| \leqslant M|x| \tag{7.1}$$

for all x in X. Let $\delta_1$ and $\delta_2$ be the moduli of convexity of $(X, |\cdot|)$ and $(X, \|\cdot\|)$ respectively.

For positive $\varepsilon$ and $\mu$, choose points x and y in the unit ball of $(X, \|\cdot\|)$ such that $\|x-y\| \geqslant \varepsilon$ and $\|(x+y)/2\| \geqslant (1-\mu)(1-\delta_2(\varepsilon))$. Since $|x| \leqslant 1/m$, $|y| \leqslant 1/m$ and $|x-y| \geqslant \varepsilon/M$, we also have

$$|(x+y)/2| \leqslant (1-\delta_1(m\varepsilon/M))/m.$$

It follows that

$$(1-\mu)(1-\delta_2(\varepsilon)) \leqslant k(1-\delta_1(\varepsilon/k)),$$

where $k = M/m$. Since this inequality holds for all positive $\mu$, we conclude that

$$\delta_2(\varepsilon) \geqslant 1 - k(1-\delta_1(\varepsilon/k)). \tag{7.2}$$

Let $b > 1$ be the unique solution of the equation

$$b(1 - \delta_1(1/b)) = 1 . \tag{7.3}$$

If $1 \leqslant k < b$, then $k(1-\delta_1(1/k)) < 1$ and $\delta_2(1) > 0$ by (7.2). Hence the characteristic of convexity of $(X, \|\cdot\|)$ is also smaller than 1. Corollary 7.2 now shows that every bounded closed convex subset of $(X, \|\cdot\|)$ has the FPP.

Thus we can say that the FPP for bounded closed convex subsets of spaces with $\varepsilon_0 < 1$ is stable in the sense that it remains intact when the norm is slightly changed.

As a matter of fact, we can rephrase our result in the following way. Recall that the Banach-Mazur distance coefficient of X with respect to a Banach space Y is

$$d(X,Y) = \inf\{\|S\|\ \|S^{-1}\| : S : X \to Y \text{ is an isomorphism}\}.$$

Theorem 7.3. Let X be a Banach space with $\varepsilon_0 < 1$, and let $b > 1$ satisfy (7.3). Let Y be another Banach space. If $d(X,Y) < b$, then every bounded closed convex subset of Y has the FPP.

Let $\gamma(x)$ be the supremum of those numbers which can replace b in Theorem 7.3. For a Hilbert space H, the solution to (7.3) is $\sqrt{5}/2$. Hence $\gamma(H) \geq \sqrt{5}/2$. We shall improve this estimate in the next section.

## 8. UNIFORMLY LIPSCHITZIAN MAPPINGS

Let C be a bounded closed convex subset of a Banach space $(X, |\cdot|)$. Once again, let X be renormed by the equivalent norm which satisfies (7.1) , and let $k = M/m$. If $T : C \to C$ is nonexpansive with respect to $\|\cdot\|$, then for all $n \geq 1$, $|T^n x - T^n y| \leq \|T^n x - T^n y\|/m \leq \|x-y\|/m \leq k|x-y|$, where $k = M/m$. Thus all iterates of T are Lipschitzian with respect to $|\cdot|$, and their Lipschitz constants are less than or equal to k. In fact, the same conclusion can be drawn if T is nonexpansive with respect to any equivalent metric on C. That is, if $d(Tx,Ty) \leq d(x,y)$ for a metric d on C (not necessarily induced by a norm) satisfying $m|x-y| \leq d(x,y) \leq M|x-y|$ for all x and y in C.

We shall say that $T : C \to C$ is uniformly Lipschitzian if there is a constant k such that

$$|T^n x - T^n y| \leq k|x-y|$$

for all $n = 1,2,\ldots$ and for all $x$ and $y$ in $C$. Given such a mapping $T$, we can define an equivalent metric $d$ on $C$ by

$$d(x,y) = \sup\{|T^n x - T^n y|: \; n = 0,1,2\ldots\}.$$

Clearly $|x-y| \leqslant d(x,y) \leqslant k|x-y|$ and $d(Tx,Ty) \leqslant d(x,v)$.

In other words, a mapping is uniformly Lipschitzian if and only if it is nonexpansive with respect to some equivalent metric. It is natural, therefore, to study the existence of fixed points of such mappings. To this end, assume that $\varepsilon_0(X) < 1$, and consider the intersection $D$ of the two closed balls $\overline{B}(x,(1+\mu)r)$ and $\overline{B}(y,k(1+\mu)r)$, where $\mu > 0$ and $k > 1$. Assume further that $|x-y| \geqslant (1-\mu)r$ and let $z = (x+y)/2$. If $u$ belongs to $D$, then

$$|u-x| \leqslant (1+\mu)r, \quad |u-y| \leqslant k(1+\mu)r$$

and so

$$|u-z| \leqslant \left(1 - \delta\left(\frac{1-\mu}{k(1+\mu)}\right)\right)k(1+\mu)r \leqslant \alpha r$$

with $\alpha < 1$ if $\mu$ is sufficiently small and $k$ is sufficiently close to 1. In other words, if $\varepsilon_0(X) < 1$, then there is a number $c > 1$ with the following property: For any $k < c$ there are positive numbers $\mu$ and $\alpha < 1$ such that for any $x,y$ in $X$ and $r > 0$ with $|x-y| \geqslant (1-\mu)r$ the set $\overline{B}(x,(1+\mu)r) \cap \overline{B}(y,k(1+\mu)r)$ is contained in a closed ball of radius $\alpha r$.

Let $\kappa(X)$ be the supremum of those numbers which can replace $c$ in the last sentence. We have shown that if $\varepsilon_0(X) < 1$ and $b$ is the solution of equation (7.3), then $\kappa(x) \geqslant b > 1$. (The converse is also true; if $\kappa(X) > 1$, then $\varepsilon_0(X) < 1$ [28].) It is not difficult to see that for a Hilbert space $H$, $\kappa(H) = \sqrt{2}$. The relationship between $\kappa(X)$ and uniformly Lipschitzian mappings is given by the following theorem, due to E.A. Lifschitz [62].

Theorem 8.1. Let C be a bounded closed convex subset of a Banach space with $\kappa(X) > 1$, and let $T : C \to C$ be a uniformly Lipschitzian mapping with Lipschitz constant k. If $k < \kappa(X)$, then T has a fixed point.

Proof. For any point x in C, let $r(x) = \inf\{r > 0:$ there is $y \in C$ such that $|x-T^n y| \leq r$ for $n=1,2,\ldots\}$. Let $\mu$ be the positive number associated with $k < \kappa(X)$. Given x in C, there is an m such that $|x-T^m x| \geq (1-\mu)r(x)$, and a point y in C for which

$$|x-T^n y| \leq (1+\mu)r(x) \quad \text{for all } n \geq 0.$$

The set $D = \overline{B}(x,(1+\mu)r(x)) \cap \overline{B}(T^m x,k(1+\mu)r(x))$ is contained in a closed ball of center z and radius $\alpha r(x)$. If $n > m$, then $|T^m x - T^n y| \leq k|x - T^{n-m}y| \leq k(1+\mu)r(x)$, so that $\{T^n y : n > m\}$ is contained in D and therefore in $\overline{B}(z,\alpha r(x))$. Hence $r(z) \leq \alpha r(x)$. For u in D we also have

$$|z-x| \leq |z-u| + |u-x| \leq \alpha r(x) + (1+\mu)r(x),$$

so that $|z-x| \leq Ar(x)$ for some constant A. It follows that we can construct a sequence of points $\{x_n\}$ in C such that $r(x_{n+1}) \leq \alpha r(x_n)$ and $|x_{n+1} - x_n| \leq Ar(x_n)$. Since $\alpha < 1$, $\{x_n\}$ is Cauchy and converges to a point w in C. This point must be a fixed point of T because $r(w) = 0$.

Let $\beta(X)$ be the supremum of those numbers which can replace $\kappa(X)$ in Theorem 8.1. We have

$$\gamma(X) \geq \beta(X) \geq \kappa(X).$$

As we already know, $\kappa(H) = \sqrt{2}$ for a Hilbert space H. Hence $\gamma(H) \geq \sqrt{2}$, which improves the estimate obtained in Section 7.

Although $\gamma(H)$ may well be infinite, $\beta(X)$ is finite for all infinite-dimensional spaces X. To see this, let B and S be the closed unit ball and unit sphere of $X$ respectively. It is known [9,71] that there is a Lipschitzian retraction $R$ of $B$ onto $S$. As a matter of fact, the proof in [9] shows that there is a constant $q$ ( $\leq 64$ ) such that for any infinite-dimensional normed space $X$ there is a Lipschitzian retraction $R$ of $B$ onto $S$ with $L(R) \leq q$. In Hilbert space, $L(R) > 4.4$ for any such retraction, and there is a retraction with $L(R) \leq 16$. The mapping $T : B \to B$ defined by $Tx = -Rx$ for $x$ in $B$ is uniformly Lipschitzian and fixed point free. The first example of a fixed point free uniformly Lipschitzian mapping in $\ell^2$ (with $k=2$) was given in [39]. New lower bounds for $\beta(L^p)$, $1 < p < \infty$, $p \neq 2$, can be found in the recent papers by T.C. Lim [64]. We conclude this section with an example of J.B. Baillon [4] which shows that $\beta(H) \leq \pi/2$.

Example 8.2. Let $H = \ell^2$,

$$B_1^+ = \{x \in \ell^2 : |x| \leq 1,\ x_1 = 0 \text{ and } x_j \geq 0 \text{ for all } j \geq 2\},$$

$$S^+ = \{x \in \ell^2 : |x| = 1,\ x_j \geq 0 \text{ for all } j \geq 1\},$$

$S_1^+ = S^+ \cap B_1^+$, and $e = (1,0,0,\ldots)$. Define $R : B_1^+ \to S^+$ by

$$Rx = \begin{cases} (\cos \frac{\pi}{2} |x|)e + (\sin \frac{\pi}{2} |x|)\, x/|x|, & x \neq 0 \\ e, & x = 0. \end{cases}$$

Note that $R$ coincides with the identity on $S_1^+$. For any two points $x \neq y$ in $B_1^+$, consider the curve $p(t) = R((1-t)x + ty)$, $0 \leq t \leq 1$. A computation shows that $|p'(t)| < (\pi/2)|x-y|$ for all $t$. Therfore $|Rx-Ry| = \int_0^1 |p'(t)|dt < (\pi/2)|x-y|$. Now let $Q$ denote the right shift $(x_1,x_2,\ldots) \to (0,x_1,x_2,\ldots)$, and consider the mapping $T : B_1^+ \to S_1^+$ defined by $T = QR$. Since $Q$ is an isometry and

$T^n = Q^nR$, T is uniformly Lipschitzian with Lipschitz constant $\pi/2$. But it is fixed point free: If $x = Tx$, then $|x| = 1$ and $x = Qx$, which is impossible.

## 9. ACCRETIVE OPERATORS

A major reason for the current interest in nonexpansive mappings is their connection with certain nonlinear differential equations. These are the evolution equations governed by accretive operators.

Let X be a Banach space. A subset A of $X \times X$ of domain D(A) and range R(A) is called accretive if for all $x_i \in D(A)$, $y_i \in Ax_i$, i = 1,2, and $r > 0$,

$$|x_1 - x_2| \leq |x_1 - x_2 + r(y_1 - y_2)|. \tag{9.1}$$

Thus the resolvent $J_r = (I+rA)^{-1} : R(I+rA) \to D(A)$ is a nonexpansive mapping for all positive r. Using Lemma 3.3, we see that A is accretive if and only if for each $x_i \in D(A)$ and each $y_i \in Ax_i$, i=1,2, there exists $j \in J(x_1-x_2)$ such that

$$(y_1 - y_2, j) \geq 0. \tag{9.2}$$

Therefore in Hilbert space the class of accretive operators coincides with the class of monotone operators.

It is well known that many phenomena and processes in the natural world can be modeled by partial differential equations and that certain partial differential equations may be interpreted as "simpler" initial value problems for ordinary differential equations in infinite-dimensional Banach spaces. Evolution equations that are governed by accretive operators occur, for example, in problems involving the heat, wave and Schrödinger equations, flows in porous media, a single conservation law, the Carleman model in gas kinetics, nonlinear diffusion, and Bellman's equation of dynamic programming. The solutions to these evolution equations give rise to (nonlinear) nonexpansive semigroups on subsets C of Banach spaces X. These are functions $S : [0,\infty) \times C \to C$ satisfying the following conditions:

$$S(t_1 + t_2)x = S(t_1)S(t_2)x \text{ for } t_1,t_2 \geq 0 \text{ and } x \in C;$$

$$|S(t)x - S(t)y| \leq |x-y| \text{ for } t \geq 0 \text{ and } x,y \in C;$$

$$S(0)x = x \text{ for } x \in C;$$

$$S(t)x \text{ is continuous in } t \geq 0 \text{ for each } x \in C.$$

It is not our purpose in this text to discuss the extensive theory of accretive operators and the nonlinear semigroups they generate. Instead, we close this section with a few examples of accretive operators in $L^p$ spaces, $1 < p < \infty$. Let $\beta$ be a monotone graph in $R^1$, and let $\Omega$ be a bounded domain in $R^n$ with smooth boundary $\partial\Omega$. With appropriate domains, the operators $A_1u = -\Delta u + \beta(u)$ with a homogeneous Neumann boundary condition, and $A_2u = -\Delta u$ with $-\partial u/\partial n \in \beta(u)$ on $\partial\Omega$, are accretive in $L^p(\Omega)$. The operator $A_3u = -\sum_{i=1}^{n} (\partial/\partial x_i)(|\partial u/\partial x_i|^{r-1} \partial u/\partial x_i)$ is also accretive for $r \geq 1$. In addition to the $L^p$ spaces, other uniformly convex function spaces (e.g. Sobolev and Orlicz spaces) arise in applications. Accretive operators occur in such spaces too. For more information on accretive and monotone operators we refer the reader to [11] and [15].

## 10. PERIODIC SOLUTIONS

Our aim in this section is to sketch a simple application of Theorem 5.2.

A monotone operator A in a Hilbert space H is said to be maximal monotone if there is no monotone operator in H which strictly contains A. It is said to be coercive if there exists $x_0$ in H such that

$$\lim_{\substack{|x| \to \infty \\ y \in Ax}} \frac{(y, x-x_0)}{|x|} = \infty .$$

It is known that if $A \subset H \times H$ is maximal monotone, then cl(D(A)), the closure of the domain of A, is convex. It is also known that for each x in cl(D(A)) and for each f in $L^1(0,T; H)$, there exists a unique weak solution to the initial value problem

$$\begin{cases} u'(t) + Au(t) \ni f(t), \ 0 < t < T, \\ u(0) = x . \end{cases} \tag{10.1}$$

These facts will be used in the proof of the following result [11].

Theorem 10.1. Let A be a maximal monotone coercive operator in a Hilbert space H. Then for each f in $L^1(0,T; H)$ there exists a weak solution to the problem

$$\begin{cases} u'(t) + Au(t) \ni f(t), \ 0 < t < T, \\ u(0) = u(T) . \end{cases} \tag{10.2}$$

Proof. Define a self-mapping S of cl(D(A)) in the following way: For each x in cl(D(A)), let $u : [0,T] \to \text{cl}(D(A))$ be the unique weak solution of (10.1), and set Sx = u(T). If u and v are any two weak solutions of (10.1), then the monotonicity of A implies that $|u(t) - v(t)| \le |u(s) - v(s)|$ for all $0 \le s \le t \le T$. Hence S is nonexpansive. For a fixed y in cl(D(A)), let $u_n$ be the unique weak solutions of (10.1) with $x = S^n(y)$. Then $u_n(T) = S^{n+1}(y)$, and since S is nonexpansive,

$$|u_n(0)| - |u_n(T)| = |S^n(y)| - |S^{n+1}(y)|$$

$$\leq |S^n(y) - S^{n+1}(y)| \leq |y - Sy|.$$

Lemma 3.6 of [11] and the coercivity of A now show that the sequence $\{u_n\}$ is uniformly bounded on $[0,T]$. In particular, the sequence of iterates $\{S^n(y)\}$ remains bounded as $n \to \infty$. Since $cl(D(A))$ is convex, we can now apply Theorem 5.2 to conclude that S has a fixed point. Such a fixed point is a solution of (10.2).

A related application of Theorem 5.2 to a certain parabolic boundary value problem which occurs in the testing of materials can be found in the recent paper by B. Kawohl and R. Ruhl [55].

## 11. FIRMLY NONEXPANSIVE MAPPINGS

Let D be a subset of a Banach space X. A mapping $T : D \to X$ is said to be firmly nonexpansive if for each x and y in D, the convex function $\phi : [0,1] \to [0,\infty)$ defined by

$$\phi(s) = |(1-s)x + sTx - ((1-s)y + sTy)|$$

is non-increasing. It is clear that every firmly nonexpansive mapping is nonexpansive. The following lemma is a consequence of Lemma 3.3.

<u>Lemma 11.1.</u> Let D be a subset of a real Banach space X, J the duality map of X, and T a mapping from D into X. Then the following are equivalent:

(a) $T : D \to X$ is firmly nonexpansive;

(b) For each x and y in D,

$$|Tx - Ty| \leq |(1-s)(x-y) + s(Tx-Ty)| \text{ for all } 0 \leq s \leq 1;$$

(c) For each x and y in D, there is $j \in J(Tx - Ty)$ such that

$$|Tx - Ty|^2 \leq (x - y, j);$$

(d) For each x and y in D,

$$|Tx - Ty| \leq |r(x - y) + (1 - r)(Tx - Ty)| \text{ for all } r > 0.$$

We have already seen in Section 3 that the nearest point projection onto a closed convex subset of a Hilbert space is firmly nonexpansive. It is also not difficult to see that the resolvent $J_r$ of an accretive operator is firmly nonexpansive. As a matter of fact, T is firmly nonexpansive if and only if it is the resolvent $(I + A)^{-1}$ for some accretive operator $A \subset X \times X$. Any linear projection of norm 1 is also firmly nonexpansive.

In Hilbert space, the inequality in part (c) of Lemma 11.1 can be written as

$$|Tx - Ty|^2 \leq (x-y, Tx - Ty). \tag{11.1}$$

This latter inequality, in turn, is equivalent to the nonexpansiveness of $2T - I$. Thus we have the following fact.

Proposition 11.2. In Hilbert space, a mapping T is firmly nonexpansive if and only if $2T - I$ is nonexpansive.

In other words, in Hilbert space T is firmly nonexpansive if and only if it is of the form $(I + S)/2$ with a nonexpansive S.

In this connection, recall that a Banach space X is said to have property (S) [21] if there exists a constant $b > 0$ such that if $|x+ry| \geq |x|$ for all $r \geq 0$, then $|x+y| \geq |x-by|$. If X has property (S) and T is firmly nonexpansive, then $aT + (1-a)I$ is nonexpansive for $0 \leq a \leq 1+b$. Also, if S is nonexpansive, then $(1-c)I + cS$ is firmly nonexpansive for $0 \leq c \leq b/(1+b)$. If X has property (S), then it is strictly convex. On the other hand, a two-dimensional example shows that X and $X^*$ may be uniformly convex without X possessing this property. A Hilbert space has property (S) with $b = 1$. It is shown in [21] that there are non-Hilbert spaces with this property. For example, any $L^p$ space, $1 < p < \infty$, has property (S).

Let C be a closed convex subset of a Banach space X. We show now that to each nonexpansive $T : C \to C$ we can associate a family of firmly nonexpansive mappings $\{G_t : 0 \leq t < 1\}$ with the same fixed point sets. Indeed, for any point a in C and $0 \leq t < 1$, let $y_a(t)$ be the unique fixed point of the strict contraction $g_a : C \to C$ defined by $g_a(x) = (1-t)a + tTx$ for x in C. We call $y_a : [0,1) \to C$ an approximating curve for T. We now define the mappings $G_t : C \to C$ by $G_t(x) = y_x(t)$. In other words,

$$G_t(x) = (1-t)x + tTG_t(x) \text{ for all } x \text{ in } C. \tag{11.2}$$

It is clear that for each $0 \leq t < 1$, the fixed point set of $G_t$ coincides with that of T. Since (11.2) can be rewritten as

$$x = G_tx + r(G_tx - TG_tx)$$

with $r = t/(1-t)$, we see that $G_t$ is the resolvent $J_r$ of the accretive operator $A = I-T$. Hence $G_t$ is indeed firmly nonexpansive. The behavior of $G_t$ when $t \to 1$ will be discussed in the next section.

We conclude the present section with an observation concerning the localization of fixed points in a Hilbert space H.

Let $\overset{+}{Eq}(u,v)$ denote the half-space $\{w \in H: |w-v| \leq |w-u|\} = \{w \in H : (w-(v+u)/2, v-u) \geq 0\}$. Let F(T) be the fixed point set of a nonexpansive $T : H \to H$.

It is not difficult to see that

$$F(T) = \cap \{\overset{+}{Eq}(x,Tx): x \in H\}. \tag{11.3}$$

If T is firmly nonexpansive, then (11.3) can be improved. Indeed, in this case the function $|z-((1-s)x + sTx)|$ with z in F(T) is non-increasing for $0 \leq s \leq 1$.

Hence $(z-Tx, Tx - x) \geq 0$ and z belongs to $\overset{+}{Eq}(x,2Tx - x)$. It follows that (11.3) can be replaced by

$$F(T) = \bigcap \{\overset{+}{Eq}(x, 2Tx-x) : x \in H\} \tag{11.4}$$

when T is firmly nonexpansive. Note that $\overset{+}{Eq}(u,2v-u)$ is always contained in $\overset{+}{Eq}(u,v)$.

## 12. APPROXIMATING CURVES

Let C be a closed convex subset of a Banach space X, and let $T : C \to C$ be nonexpansive. In this section we wish to discuss the behavior of the mappings $G_t : C \to C$, defined in the preceding section by (11.2), as $t \to 1$.

To this end, we fix a point x in C and consider the approximating curve $y(t) = y_x(t)$. Suppose that there exists a sequence $t_n \to 1$ such that $\{y(t_n)\}$ remains bounded as $n \to \infty$. Pick a point w in C and set $R = \limsup_{n\to\infty} |w-y(t_n)|$. Let

$$K = \{z \in C : \limsup_{n\to\infty} |z-y(t_n)| \leq R\}.$$

K is a nonempty bounded closed convex subset of C. Now let z be in K. Then

$$|Tz - y(t_n)| = |Tz - (1-t_n)x - t_nTy(t_n)|$$

$$\leq (1-t_n)|z-x| + t_n|Tz-Ty(t_n)|$$

$$\leq (1-t_n)|z-x| + t_n|z-y(t_n)|,$$

so that Tz belongs to K. Thus K is invariant under T and we are led to the following result.

Theorem 12.1. Let C be a closed convex subset of a Banach space X, $T : C \to C$ a nonexpansive mapping, and $G_t : C \to C$, $0 \leq t < 1$, the family of mappings defined by (11.2). If each bounded closed convex subset of X has the FPP and T is fixed point free, then $\lim_{t \to 1} |G_t(x)| = \infty$ for each x in C.

What happens when T has a fixed point? It was shown by Halpern [46] in 1967 that in Hilbert space the strong $\lim_{t \to 1} G_t(x)$ exists for each x and is a fixed point of T. Although this result was extended to a restricted class of Banach spaces, it has remained an open question whether it is true in, say, the $L^p$ spaces, $1 < p < \infty$, $p \neq 2$. Theorem 12.2 below provides an affirmative answer to this problem. It is due to Reich [86].

Before stating this theorem we recall that a Banach space X is said to be uniformly smooth if the limit (3.1) exists uniformly for x and y in the unit sphere of X. A Banach space X is uniformly smooth if and only if its dual $X^*$ is uniformly convex. It is shown in [96] that a uniformly smooth space has normal structure. Since such a space is also reflexive, each bounded closed convex subset of it has the FPP by Theorem 6.4. This result, originally due to Baillon [4], will be used in the proof of Theorem 12.2. We also recall that a Banach limit LIM is a bounded linear functional on $\ell^\infty$ of norm 1 such that

$$\liminf_{n \to \infty} x_n \leq \mathrm{LIM}\{x_n\} \leq \limsup_{n \to \infty} x_n$$

and

$$\mathrm{LIM}\{x_n\} = \mathrm{LIM}\,\{x_{n+1}\} \text{ for all } \{x_n\} \text{ in } \ell^\infty.$$

<u>Theorem 12.2.</u> Let C be a closed convex subset of a uniformly smooth Banach space, $T : C \to C$ a nonexpansive mapping, and $G_t : C \to C$, $0 \leq t < 1$, the family of mappings defined by (11.2). If T has a fixed point, then for each x in C the strong $\lim_{t \to 1} G_t(x)$ exists and is a fixed point of T.

<u>Proof.</u> Fix a point x in C, denote $G_t(x)$ by $y(t)$, and let w be a fixed point of T. Since

$$y(t) - w = (1-t)(x-w) + t(Ty(t) - Tw),$$

$|y(t) - w| \leq |x-w|$ and $\{y(t)\}$ remains bounded as $t \to 1$. We also have $\lim_{t\to 1}(y(t) - Ty(t)) = 0$. Now let $t_n \to 1$ and $y_n = y(t_n)$. Let LIM be a Banach limit and define $f : C \to [0,\infty)$ by

$$f(z) = \mathrm{LIM}\ \{|y_n - z|^2\}.$$

Since f is continuous and convex, $f(z) \to \infty$ as $|z| \to \infty$, and X is reflexive, f attains its infimum over C. Let K be the set of minimizers of f. If $u \in C$, then

$$f(Tu) = \mathrm{LIM}\{|y_n - Tu|^2\} = \mathrm{LIM}\{|Ty_n - Tu|^2\} \leq \mathrm{LIM}\{|y_n - u|^2\} = f(u).$$

Therefore K is invariant under T. Since it is also bounded, closed and convex, it must contain a fixed point of T by the discussion preceding the statement of Theorem 12.2. Denote such a fixed point by v, and let $J : X \to X^*$ be the duality map of X. Since v is a fixed point of T, $(y_n - Ty_n, J(y_n - v)) \geq 0$ for all n. It follows that

$$(y_n - x, J(y_n - v)) \leq 0 \tag{12.1}$$

for all n. On the other hand, we note that if $0 < t \leq 1$ and $z \in C$, then

$$|y_n - v|^2 = |y_n - (1-t)v - tz + t(z-v)|^2$$

$$\geq |y_n - (1-t)v - tz|^2 + 2t(z-v, J(y_n - (1-t)v - tz)).$$

Let $\varepsilon > 0$ be given. The uniform smoothness of X implies that $|(z-v, J(y_n-v) - J(y_n-(1-t)v-tz)| < \varepsilon$ for all small enough t. Consequently,

$$(z-v, J(y_n - v)) < \varepsilon + (z-v, J(y_n-(1-t)v-tz))$$

$$\leq \varepsilon + \frac{1}{2t}(|y_n - v|^2 - |y_n-(1-t)v-tz|^2) .$$

Since v is in K, it follows that

$$\text{LIM}\{(z-v, J(y_n - v))\} \leq 0 \tag{12.2}$$

for all z in C. Choosing z = x and combining (12.2) with (12.1), we conclude that $\text{LIM}\{|y_n - v|^2\} \leq 0$. Therefore there is a subsequence of $\{y_n\}$ which converges strongly to v. To complete the proof, suppose that $\lim_{k\to\infty} y(t_{n_k}) = v_1$ and $\lim_{k\to\infty} y(t_{m_k}) = v_2$. Then by (12.1), $(v_1 - x, J(v_1 - v_2)) \leq 0, (v_2 - x, J(v_2 - v_1)) \leq 0$, and $v_1 = v_2$.

Although the uniform smoothness assumption can be weakened, Theorem 12.2 cannot be extended to all Banach spaces. To see this, let $C = X = C[0,1]$ and define $T : C \to C$ by $(Tx)(s) = sx(s)$, $0 \leq s \leq 1$.

Theorem 12.2 is of special interest because it leads to a strong convergence result for the explicit iteration

$$x_{n+1} = (1-k_n)x_0 + k_n Tx_n . \tag{12.3}$$

The Hilbert space case is again due to Halpern [46].

<u>Corollary 12.3.</u> In the setting of Theorem 12.2, let $\{x_n\}$ be defined by (12.3) with $k_n = 1 - n^{-a}$, where $0 < a < 1$. Then $\{x_n\}$ converges strongly to a fixed point of T.

Combining Theorem 12.2 with a theorem of Haydon, Odell and Sternfeld [50] we also obtain the following fixed point result.

<u>Corollary 12.4.</u> Let $C_1$ be a closed subset of a Banach space $E_1$, and let $C_2$ be a bounded closed convex subset of a uniformly

smooth Banach space $E_2$. If $C_1$ has the FPP, then so does the subset $C_1 \oplus C_2$ of $(E_1 \oplus E_2)_\infty$.

For more results in this direction, see the paper by W.A. Kirk and Y. Sternfeld [59].

## 13. NONEXPANSIVE RETRACTIONS

What is the limit obtained in Theorem 12.2? It is certainly a firmly nonexpansive retraction of C onto the fixed point set F(T) of T. Denote this retraction by Q. Part (c) of Lemma 11.1 shows that

$$(x - Qx,\ J(y-Qx)) \leq 0 \tag{13.1}$$

for all x in C and y in F(T). The following lemma will show that Q is the unique sunny nonexpansive retraction of C onto F(T). (Recall that a retraction is said to be sunny if it satisfies (3.7).) (A sunny nonexpansive retraction is necessarily firmly nonexpansive.)

<u>Lemma 13.1.</u> Let C be a closed convex subset of a smooth Banach space X. Let $J : X \to X^*$ be the duality map of X, K a subset of C, and $Q : C \to K$ a retraction. Then the following are equivalent:

(a) $(x-Qx, J(y-Qx)) \leq 0$ for all x in C and y in K;

(b) $|Qz-Qw|^2 \leq (z-w, J(Qz-Qw))$ for all z and w in C;

(c) Q is both sunny and nonexpansive.

Hence, there is at most one sunny nonexpansive retraction on K.

<u>Proof.</u> Suppose (a) holds and let z and w belong to C. Set $j = J(Qz - Qw)$. Since both $(w-Qw,j)$ and $(z-Qz, -j)$ are nonpositive, $(Qz-Qw + w-z,\ j) \leq 0$, and (b) follows. Assume now that (b) holds and let $x \in C$ and $y \in K$. Inserting $z = y = Qy$ and $w = x$ in (b)

we obtain $(y-x, J(y-Qx)) \geq |y-Qx|^2 = (y-Qx, J(y-Qx))$ and (a) follows. Suppose $Q$ is both sunny and nonexpansive. Let $x \in C$, $y \in K$ and put $Qx = v$. The set $D = \{v + t(x-v): t \geq 0\}$ is convex. If $w \in D$, then $|y-v| = |Qy - Qw| \leq |y-w|$. Part (d) of Proposition 3.4 now implies that $(x - v, J(y-v)) \leq 0$ and (a) follows. Conversely, suppose (a) holds. $Q$ is nonexpansive by (b). Let $Qx = v$, $t \geq 0$, $w = v + t(x-v)$, and $j = J(Qw-v)$. By (a), both $(x-v,j)$ and $(w-Qw, -j)$ are nonpositive. But $t(v-x) = v-w$. Hence $(Qw-v,j) \leq 0$ and $Qw = v$. Finally suppose that both $P$ and $Q$ are sunny nonexpansive retractions. Let $x \in C$ and $j = J(Px-Qx)$. By (a), $(x-Qx, j)$ and $(x-Px, -j)$ are nonpositive. Hence $(Px-Qx, j) \leq 0$ and $Px = Qx$.

Recall that the nearest point projection $P$ of a Hilbert space $H$ onto a closed convex subset $K$ of $H$ is characterized by $(y-Px, x-Px) \leq 0$, where $x \in H$ and $y \in K$. This inequality can be extended to smooth Banach spaces in two ways: Either

$$(y-Px, J(x-Px)) \leq 0, \text{ or } (x-Px, J(y-Px)) \leq 0.$$

The first possibility yields nearest point projections (see Section 3). We have just seen that the second yields sunny nonexpansive retractions. (Outside Hilbert space, nearest point projections are still sunny, but they are no longer nonexpansive.)

Combining our present discussion with Theorem 12.2 we obtain the following result (first proved in [18] by a different method). We say that a subset $D$ of $C$ is a (sunny) nonexpansive retract of $C$ if either $D$ is empty or there exists a (sunny) nonexpansive retraction of $C$ onto $D$.

Theorem 13.2. Let $C$ be a closed convex subset of a uniformly smooth Banach space. If $T : C \to C$ is nonexpansive, then the fixed point set of $T$ is a sunny nonexpansive retract of $C$.

In this connection we also quote the following result of Bruck [17].

Theorem 13.3. Let C be a closed convex subset of a reflexive Banach space every bounded closed convex subset of which has the FPP. If $T : C \to C$ is nonexpansive, then its fixed point set is a nonexpansive retract of C.

Recall that is is still an open question whether bounded closed convex subsets of reflexive Banach spaces have the FPP. For a weak sufficient condition on the underlying space which guarantees that nonexpansive retracts are, in fact, sunny nonexpansive retracts see [85].

It is known [81] that every closed convex subset of a Banach space X is a nonexpansive retract of X if and only if X is either two-dimensional or (isometric to) a Hilbert space. Therefore Theorems 13.2 and 13.3 show that if X is a non-Hilbert space with $\dim(X) \geq 3$, then not every closed convex subset of X is the fixed point set of a nonexpansive $T : X \to X$.

Here are some examples of nonexpansive retracts in $L^p$ spaces. A closed linear subspace of $L^p$ is a nonexpansive retract of the space if and only if it is isometric to another $L^p$ space. The set $\{f \in L^p(\Omega): |f(x)| \leq 1 \text{ a.e. in } \Omega\}$ and the positive cone in $L^p$ are nonexpansive retracts of $L^p$. The set

$$\{f \in H_0^1(\Omega): |\nabla f(x)| \leq 1 \text{ a.e. in } \Omega\}$$

is also a nonexpansive retract of $L^p(\Omega)$, $p \geq 2$. Although in smooth Banach spaces sunny nonexpansive retractions onto arbitrary (closed) subsets are unique, nonexpansive retractions are not unique in general, even in Hilbert space. To see this, let H be a Hilbert space and consider, for example, the sets

$$K = \{x \in H : |x-a| \leq r\} \text{ and } B = \{x \in H : |x| \leq |a| + r\},$$

where $a \neq 0$ and $r > 0$. Let $P_K$ and $P_B$ denote the nearest point projections onto K and B, respectively. Then $P_K P_B$ is a nonexpansive retraction of H onto K which is different from $P_K$.

On the other hand, in smooth Banach spaces nonexpansive retractions onto linear (closed) subspaces are unique. To see this, let K be a linear subspace of a smooth Banach space X, and let $P : X \to K$ be a nonexpansive retraction. If u belongs to K, then for each x in X, $|x-tu|^2 + 2(Px-x, J(x-tu)) \leq |Px-tu|^2 \leq |x-tu|^2$ for all $-\infty < t < \infty$. Dividing by t and then letting $t \to \pm\infty$, we obtain that $(Px-x, J(u)) = 0$ for all u in K. Thus P is unique (and in fact, linear). This uniqueness result is no longer true if X is not smooth. For example, if $X = \ell_2^1$ and $K = \{(t,0) : -\infty < t < \infty\}$, then both $R_1(a,b) = (a,0)$ and $R_2(a,b) = (a+b,0)$ are sunny nonexpansive retractions of X onto K.

## 14. THE MEAN ERGODIC THEOREM

It is remarkable that the ideas of the proof of Theorem 12.2 also lead to a simple proof [22] of the nonlinear mean ergodic theorem in Hilbert space [2,19,82]. Following Lorentz [65], we say that a sequence $\{x_n\}$ in a Banach space X is weakly almost convergent to $z \in X$ if the weak $\lim_{n\to\infty} \left(\sum_{i=0}^{n-1} x_{k+i}\right)/n = z$ uniformly in $k \geq 0$.

Theorem 14.1. Let C be a closed convex subset of a Hilbert space H, $T : C \to C$ a nonexpansive mapping with a fixed point, and $x \in C$. Then $\{T^n x\}$ is weakly almost convergent to a fixed point of T which is the asymptotic center of $\{T^n x\}$.

Proof. Since C is a nonexpansive retract of H, we may assume that T is defined on all of H. Let LIM be any Banach limit, and define $f : H \to [0,\infty)$ by $f(z) = \mathrm{LIM}\{|T^n x - z|^2\}$. The set K of minimizers of f is nonempty, bounded, closed and convex. If $u \in H$, then

$$\begin{aligned} f(Tu) &= \mathrm{LIM}\{|T^n x-Tu|^2\} \\ &= \mathrm{LIM}\{|T^{n+1}x-Tu|^2\} \leq \mathrm{LIM}\{|T^n x-u|^2\} = f(u), \end{aligned}$$

so that f is a Liapunov function for T. (Here we have used the fact that $\mathrm{LIM}\{t_n\} = \mathrm{LIM}\{t_{n+1}\}$.) Therefore K is invariant under T and contains a fixed point v of T. Since on the fixed point set of T, f is independent of the particular LIM chosen, we may assume that the same v minimizes f for any LIM. It follows that

$$\mathrm{LIM}\{(z,\ T^n x - v)\} = 0$$

for all z in H and any LIM. Thus $\{T^n x - v\}$ is weakly almost convergent to 0 [65]. In other words, $\{T^n x\}$ is weakly almost convergent to v. It is also clear that v is indeed the asymptotic center of $\{T^n x\}$.

Different arguments [3,20,84] show that Theorem 14.1 is true in all uniformly convex spaces which are also uniformly smooth, except that outside Hilbert space the limit obtained is not necessarily the asymptotic center of $\{T^n x\}$ (see the $\ell_3^p$ example in Section 4.) The method of proof of Theorem 14.1 yields "dual" ergodic theorems in Banach spaces [22]. We conclude with a corollary of the mean ergodic theorem in Banach spaces.

<u>Corollary 14.2.</u> Let C be a closed convex subset of a Banach space which is both uniformly convex and uniformly smooth, $T : C \to C$ a nonexpansive mapping with a fixed point, and x a point in C. Then $\{T^n x\}$ converges weakly to a fixed point of T if and only if the weak $\lim_{n\to\infty} (T^n x - T^{n+1} x) = 0$.

<u>Proof.</u> $\lim_{n\to\infty} (x_n - x_{n+1}) = 0$ is a Tauberian condition for almost convergence.

## 15. ITERATIONS

In general, the iterates $\{T^n x\}$ of a nonexpansive mapping T with a fixed point do not converge either weakly or strongly. However, we do have positive results for firmly nonexpansive mappings.

Theorem 15.1. Let C be a closed convex subset of a Banach space X and let $T : C \to C$ be a firmly nonexpansive mapping with a fixed point. If both X and its dual $X^*$ are uniformly convex, then for each x in C, the sequence of iterates $\{T^n x\}$ converges weakly to a fixed point of T.

Proof. Denote $\{T^n x\}$ by $\{x_n\}$. By Corollary 14.2 it suffices to show that the weak $\lim_{n\to\infty} (x_n - x_{n+1}) = 0$. In fact, we shall show that the strong $\lim_{n\to\infty} (x_n - x_{n+1}) = 0$. To this end, let y be a fixed point of T. Then $d = \lim_{n\to\infty} |x_n - y|$ exists. Since T is firmly nonexpansive, $|Tx_n - y| \leq |(x_n + Tx_n)/2 - y| \leq |x_n - y|$ for all n. Hence $|(x_n + Tx_n)/2-y| \to d$ and $x_n - x_{n+1} = x_n - Tx_n \to 0$ by uniform convexity.

Instead of using Corollary 14.2, we can appeal to the Proposition in [83]. There it is shown that $\lim_{n\to\infty} (T^n x, J(f_1-f_2))$ exists for all $f_1$ and $f_2$ in the fixed point set F(T) of T. Since $x_n - Tx_n \to 0$, every subsequential weak limit of $\{x_n\}$ is a fixed point of T [14]. Let $f_1$ and $f_2$ be two such limits. Then

$$(f_1, J(f_1-f_2)) = (f_2, J(f_1-f_2)),$$

so that $f_1 = f_2$. Hence $\{x_n\} = \{T^n x\}$ converges weakly. In any case, Theorem 15.1 improves upon the original result of Opial [72]. His approach worked for the sequence spaces $\ell^p$, but not for the Lebesgue spaces $L^p$, $1 < p < \infty$.

Theorem 15.1 is not true in all Banach spaces (consider $T : C[0,1] \to C[0,1]$ defined by $(Tf)(t) = tf(t)$, $0 \leq t \leq 1$). Nor can the conclusion of weak convergence be replaced by strong convergence [37]. We do have, however, the following result [21].

Theorem 15.2. Let T be a firmly nonexpansive self-mapping of a closed convex subset C of a uniformly convex Banach space. If C

$= -C$ and T is odd, then $\{T^n x\}$ converges strongly to a fixed point of T.

What happens if a firmly nonexpansive mapping is fixed point free? Again we quote a result from [21].

Theorem 15.3. Let T be a firmly nonexpansive self-mapping of a closed convex subset C of a uniformly convex Banach space. Then T is fixed point free if and only $\lim_{n\to\infty} |T^n x| = \infty$ for all x in C.

Theorem 15.3 is not true if T is merely nonexpansive, even in Hilbert space [31].

By Proposition 11.2, in Hilbert space T is firmly nonexpansive if and only if it is of the form $(I+S)/2$ with a nonexpansive S. It is of interest to note that Theorems 15.1, 15.2 and 15.3 remain true for such averaged mappings even outside Hilbert space [5].

We now turn to arbitrary nonexpansive mappings and present two new results due to Plant and Reich [74] which are of interest when T is fixed point free. First we recall that the norm of a Banach space X is said to be Fréchet differentiable if for each x in the unit sphere $U = \{x \in X : |x| = 1\}$ of X the limit (3.1) is attained uniformly for y in U. Obviously this is a weaker assumption than uniform smoothness (equivalently, uniform convexity of the dual $X^*$). As a matter of fact, Corollary 14.2 and Theorem 15.1 remain true if the uniform smoothness hypothesis there is replaced by this weaker assumption. The first result we quote is of interest because in contrast with previous results in this direction, the domain D of T is not assumed to be convex, nor is $A = I-T$ assumed to satisfy the range condition $R(I+rA) \supset D$ for all $r > 0$.

Theorem 15.4. Let D be a closed subset of a Banach space X and $T : D \to D$ a nonexpansive mapping. If $X^*$ has a Fréchet differentiable norm, then w = the strong $\lim_{n\to\infty} T^n x/n$ exists for each x in D (and is independent of x).

It is known [60] that the convergence of $\{T^n x/n\}$ for all nonexpansive $T : X \to X$ is actually equivalent to the Fréchet differentiability of the norm of $X^*$.

There are circumstances in which it is possible to identify the limit of Theorem 15.4.

Theorem 15.5. In the setting of Theorem 15.4 assume, in addition, that $X^*$ is strictly convex. Then the following are equivalent:

(a) $\liminf_{t\to\infty} d(0,R(I+tA))/t = 0$;

(b) $-w$ is the unique point of least norm in $cl(R(A))$;

(c) $|w| = d(0,R(A))$.

Corollary 15.6. Let C be a closed convex subset of a Banach space X, $T : C \to C$ a nonexpansive mapping, and x a point in C. Assume that X is smooth and that $X^*$ has a Fréchet differentiable norm. Then the strong $\lim_{n\to\infty} T^n x/n = -v$, where v is the point of least norm in $cl(R(I-T))$.

Proof. Since C is convex, Banach's fixed point theorem shows that $A = I-T$ does satisfy the range condition.

Finally, we note that if X is uniformly convex and T is firmly nonexpansive, then $\lim_{n\to\infty} (T^n x - T^{n+1} x) = v$ [87].

## 16. INWARD MAPPINGS

For x in C, a closed convex subset of a Banach space X, let

$$I_C(x) = \{z \in X : z = x + a(y-x) \text{ for some } y \in C \text{ and } a \geq 0\}.$$

A mapping $f : C \to E$ is said to be weakly inward if f(x) belongs to the closure of $I_C(x)$ for each x in C. This concept is closely related to invariance criteria for solutions of differential equations. In recent years there has also been great interest in fixed point theorems for mappings which do not map their domains of definition into themselves, but instead satisfy inwardness conditions. As an example, we shall now sketch a proof of the following result [79].

Theorem 16.1. Let C, a bounded closed convex subset of a Banach space X, possess the FPP. If a nonexpansive $T : C \to X$ is weakly inward, then it has a fixed point.

Proof. Let $0 < t < 1$ and $r = t/(1-t)$. The resolvent $R = (I+r(I-T))^{-1}$ is single-valued and nonexpansive on its domain D. Let z belong to C and define $g : C \to X$ by $g(x) = tTx + (1-t)z$. It is not difficult to see that the strict contraction g is also weakly inward. Therefore it can be shown that if g is fixed point free, then there is a function $f : C \to C$ and a positive p such that

$$|x-f(x)| < |x-g(x)|/p - |f(x) - g(f(x))|/p \tag{16.1}$$

for all x in C. This mapping f must have a fixed point by the Caristi-Ekeland fixed point theorem [25,32]. This, however, contradicts (16.1). Hence g does have a fixed point. It follows that C is contained in D. Every fixed point of $R : C \to C$ is also a fixed point of T.

If C has a nonempty interior, we say that $f : C \to X$ satisfies the Leray-Schauder condition if there is a point w in the interior of C such that

$$f(y) - w \neq m(y-w) \tag{16.2}$$

for all y in the boundary of C and $m > 1$. If f is weakly inward, then it satisfies the Leray-Schauder condition. It is still an open question whether Theorem 16.1 remains true when T is assumed to satisfy only the Leray-Schauder condition. We do have, however, the following result.

Theorem 16.2. Let C, a bounded closed convex subset of a uniformly convex Banach space X, have a nonempty interior. If a nonexpansive $T : C \to X$ satisfies the Leray-Schauder condition, then it has a fixed point.

Proof. For each $0 \leq t < 1$, the strict contraction $S:C \to X$ defined by $Sx = (1-t)w + tTx$, $x \in C$, satisfies the Leray-Schauder condition and therefore has a fixed point [78]. It follows that there is a sequence $\{x_n\} \subset C$ such that $\lim_{n\to\infty} (x_n - Tx_n) = 0$. By [14], every weak subsequential limit of $\{x_n\}$ is a fixed point of T.

Note that since T is not defined on all of X, we cannot use asymptotic centers to deduce the existence of a fixed point of T.

## 17. THE ALMOST FIXED POINT PROPERTY FOR NONEXPANSIVE MAPPINGS

A closed convex subset C of a Banach space is said to have the almost fixed point property for nonexpansive mappings (AFPP) if

$$\inf \{|x - Tx| : x \in C\} = 0$$

for all nonexpansive $T : C \to C$ . As we have already seen (Proposition 1.4), any bounded C has the property. As we shall see shortly, there are also unbounded C with the AFPP. As a matter of fact, the purpose of this section is to present a characterization [88] of those closed convex subsets of reflexive Banach spaces X which possess the AFPP. The best previous result was obtained by Baillon and Ray [6] who assumed that X belongs to a special class of superreflexive spaces. See also [40] (where $X = \ell^2$) and [76] (where $X = \ell^p$, $1 < p < \infty$). The set C is called linearly bounded if it has a bounded intersection with all lines in X. For example, let $X = \ell^2$ and consider the set $C = \{x \in \ell^2 : |x_n| \leq p_n\}$. This set is always linearly bounded. It is bounded if and only if $\sum_{n=1}^{\infty} p_n^2 < \infty$.

Theorem 17.1. A closed convex subset of a reflexive Banach space has the AFPP if and only if it is linearly bounded.

Proof. Let C be a closed convex subset of a (real) reflexive Banach space X, and let $X^*$ be the dual of X. To show necessity, assume that $\{y+ta : 0 \leq t < \infty\} \subset C$ for some $a \neq 0$. If x is in C then $(1-1/t)x + (y+ta)/t$ belongs to C for all $t \geq 1$. Therefore we can define a mapping $S : C \to C$ by $Sx = x+a$. This mapping is nonexpansive and $|x - Sx| = |a|$ for all x in C.

Conversely, let $T : C \to C$ be any nonexpansive mapping, and denote $\inf\{|x-Tx| : x \in C\}$ by d. It is known [87] that for each x in C there is a functional $j \in X^*$ with $|j| = d$ such that $((x-T^n x)/n, j) \geq d^2$ for all $n \geq 1$. It is also known [87] that $\lim_{n\to\infty} |T^n x|/n = d$. Let a subsequence of $\{T^n x/n\}$ converge weakly to w. Clearly $|w| \leq d$. On the other hand, $|w|d = |w||j| \geq (-w,j) \geq d^2$, so that $|w| = d$. Now let y be any point in C. Since $(1-1/n)y + T^n x/n$ belongs to C for each $n \geq 1$, we see that y+w also belongs to C. Consequently, we may conclude that the points y+mw belong to C for all $m \geq 1$. If C is linearly bounded, then this fact implies that $w = 0$, so that $d = 0$ too.

Theorem 17.1 cannot be extended to all Banach spaces. To see this, let

$$X = \ell^1, \quad C = \{x \in \ell^1 : |x_n| \leq 1 \text{ for all } n\},$$

and define $T : C \to C$ by $T(x_1,x_2,\ldots) = (1,x_1,x_2,\ldots)$. Then C is linearly bounded and T is an isometry, but $\inf\{|x-Tx| : x \in C\} = 1$.

If $X$ is finite-dimensional and $C$ is linearly bounded, then $C$ is, in fact, bounded. Hence in this case either $C$ is bounded and has the FPP, or it is unbounded and does not even have the AFPP.

We also note that if $X$ is any Hilbert space and $C$ is unbounded, then $C$ does not have the FPP [77]. Finally, we note that if $C$ is a linearly bounded, closed convex subset of a reflexive Banach space $X$, and a nonexpansive $T : C \to E$ is weakly inward, then $\inf\{|x-Tx| : x \in C\} = 0$ too. This is true because the proof of Theorem 16.1 shows that Theorem 17.1 can be applied to the resolvent $(I+r(I-T))^{-1} : C \to C$, where $I$ denotes the identity operator and $r$ is positive.

# 2

# HYPERBOLIC GEOMETRY

## 1. THE SCHWARZ LEMMA

Let $D$ be the open unit disc $\{z \in C : |z| < 1\}$ in the complex plane $C$, $\partial D$ its boundary $\{z \in C : |z| = 1\}$, and $\overline{D} = D \cup \partial D$. All holomorphic mappings of $D$ into $C$ are subject to a maximum principle which for our purpose may be stated in the following (weak) form.

Theorem 1.1. If $f : D \to C$ is holomorphic, then for any $0 < r < 1$ and any $z \in D$ with $|z| \leq r$,

$$|f(z)| \leq \max\{|f(\xi)| : |\xi| = r\}.$$

If equality holds at one point $z \in D$ with $|z| < r$, then $f$ is constant.

We denote by $\mathrm{Hol}(D)$ the class of all holomorphic mappings of $D$ into itself and by $\overline{\mathrm{Hol}}(D)$ the family of all holomorphic mappings of $D$ into $\overline{D}$. In view of the maximum principle, $\overline{\mathrm{Hol}(D)} \setminus \mathrm{Hol}(D)$ contains only unimodular constant functions.

The following result is usually called the Schwarz Lemma.

Theorem 1.2. If $f \in \mathrm{Hol}(D)$ and $f(0) = 0$, then

(a) $|f(z)| \leq |z|$ for all $z \in D$;

(b) $|f'(0)| \leq 1$.

If equality holds in (a) for one $z \neq 0$, or if equality holds in (b), then $f(z) = \lambda z$, where $\lambda$ is a unimodular constant.

Proof. Since $f(0) = 0$, there is a holomorphic function $g: D \to C$ such that $f(z) = zg(z)$. For $|z| \leq r < 1$, we have by Theorem 1.1,

$$|g(z)| \leq \max\{|f(\xi)|/r : |\xi| = r\} \leq 1/r.$$

Letting $r \to 1$, we obtain $|g(z)| \leq 1$ for all $z \in D$ and (a) follows. The maximum principle shows that if $|g(z)| = 1$ for some $z \in D$, then g is a (unimodular) constant. Since $f'(0) = g(0)$, the proof is complete.

We shall denote by Aut(D) the class of all invertible holomorphic mappings of D onto itself. It is clear that Aut(D) is a group under composition. We intend to use the Schwarz Lemma to describe the structure of this group.

A mapping of the form $f(z) = \lambda z$, where $\lambda = e^{i\phi}$ $(\phi \in R)$ is a unimodular constant, is a rotation through the angle $\phi$. It clearly belongs to Aut(D). Conversely, if $h \in \mathrm{Aut}(D)$ and $h(0) = 0$, then h must be a rotation. To see this, observe that

$$|z| = |h^{-1}(h(z))| \leq |h(z)| \leq |z|$$

and apply Theorem 1.2. For any $a \in D$, the Möbius transformation

$$m_a(z) = (z+a)/(1 + \bar{a}z)$$

also belongs to Aut(D). This fact follows from the inequality

$$1 - |m_a(z)|^2 = (1 - |a|^2)(1 - |z|^2)/|1 + \bar{a}z|^2 > 0$$

which shows that $m_a : D \to D$, and the simple observation that $m_a^{-1} = m_{-a}$.

The group Aut(D) acts transitively on D; that is for any a and b in D, there exists h in Aut(D) such that $h(a) = b$. In fact, h may be defined by $m_b \circ m_{-a}$. Our next result describes the whole group Aut(D).

<u>Theorem 1.3.</u> Any h in Aut(D) is of the form

$$h(z) = e^{i\phi} m_a(z)$$

for some $a \in D$ and $\phi \in R$.

<u>Proof.</u> Let $h \in \text{Aut}(D)$ and let $b = h(0)$. The mapping $g = m_{-b}$ h belongs to Aut(D) and $g(0) = 0$. Therefore g is a rotation and

$$h(z) = m_b(e^{i\phi} z) = e^{i\phi} m_a(z)$$

with $a = e^{-i\phi} b$ and $\phi \in R$.

Summing up, the group Aut(D) is generated by rotations and by the Möbius transformations $m_a$, $a \in D$. If the rotation through the angle $\phi$ is denoted by $r_\phi$ , then we have the following rules of composition:

$$m_a \circ r_\phi = r_\phi \circ m_{r_{-\phi}(a)}$$

for any $\phi \in R$, and

$$m_a \circ m_b = r_\phi \circ m_{m_b(a)}$$

with $\phi = 2 \arg(1 + a\bar{b})$.

We conclude this section with a consequence of the Schwarz Lemma. This result is usually referred to as the Schwarz-Pick Lemma.

Theorem 1.4. If $f \in \mathrm{Hol}(D)$, then

(a) $$\left|\frac{f(z) - f(w)}{1 - f(z)\overline{f(w)}}\right| \leq \left|\frac{z-w}{1 - z\bar{w}}\right|$$

for all z and w in D;

(b) $$\frac{|f'(w)|}{1 - |f(w)|^2} \leq \frac{1}{1 - |w|^2}$$

for all $w \in D$;

(c) If $f \in \mathrm{Aut}(D)$, then the inequalities (a) and (b) become equalities;

(d) If equality holds in (a) for one pair of points $z \neq w$, then $f \in \mathrm{Aut}(D)$;

(e) If equality holds in (b) for one $w \in D$, then $f \in \mathrm{Aut}(D)$.

Proof. Consider the automorphisms $m_w$ and $m_{-f(w)}$. The mapping $g = m_{-f(w)} \circ f \circ m_w$ belongs to Hol(D) and $g(0) = 0$. By the Schwarz Lemma, $|g(\xi)| \leq |\xi|$ for all $\xi \in D$. Setting $\xi = m_{-w}(z)$, we obtain (a). Part (b) follows from (a). If $f \in \mathrm{Aut}(D)$, then g is a rotation and (c) follows. If equality holds in (a) for some $z \neq w$, then $|g(\xi)| = |\xi|$ for $\xi = m_{-w}(z)$, g is again a rotation, and (d) follows. Finally, since $|g'(0)| = |f'(w)|(1 - |w|^2)/(1 - |f(w)|^2)$, part (e) results from part (b) of the Schwarz Lemma.

We observe that part (b) of the Schwarz-Pick Lemma shows that for any $f \in \mathrm{Hol}(D)$ and $z \in D$,

$$|f'(z)| \leq \frac{1}{1 - |z|^2} .$$

More precisely,

$$\sup\{|f'(z)|: f \in \mathrm{Hol}(D)\} = \frac{1}{1- |z|^2}$$

with equality attained for $f = m_{-z}$. This observation and the

Schwarz-Pick Lemma itself motivate our discussion of the Poincaré metric in the next section.

## 2. THE POINCARÉ METRIC

Let the curve $\gamma:[0,1] \to D$ join the points z and w in D. Assume that $\gamma'(t)$ is piecewise continuous, and let

$$L_\gamma = \int_0^1 \frac{|\gamma'(t)|dt}{1-|\gamma(t)|^2} .$$

The Poincaré distance between z and w is defined by

$$\rho(z,w) = \inf\{L_\gamma\} ,$$

where the infimum is taken over all such curves $\gamma$. It is not difficult to see that $\rho$ is in fact a metric. It has the following important property.

Theorem 2.1. If $f \in \mathrm{Hol}(D)$, then $\rho(f(z),f(w)) \leq \rho(z,w)$ for all z and w in D. If $f \in \mathrm{Aut}(D)$, then this inequality becomes an equality.

Proof. If $\gamma$ joins z and w, then $f \circ \gamma$ joins f(z) and f(w). Part (b) of the Schwarz-Pick Lemma shows that

$$L_{f\circ\gamma} = \int_0^1 \frac{|f'(\gamma(t))||\gamma'(t)|}{1 - |f(\gamma(t))|^2} dt \leq \int_0^1 \frac{|\gamma'(t)|}{1 - |\gamma(t)|^2} dt = L_\gamma,$$

and the first claim follows. If $f \in \mathrm{Aut}(D)$, then both f and $f^{-1}$ are $\rho$-nonexpansive, so that f must be an isometry.

In order to obtain an explicit formula for $\rho(z,w)$, we first evaluate $\rho(0,s)$ for $0 < s < 1$. If $\gamma$ joins 0 and s, then so does $\beta = \mathrm{Re}\,\gamma$. Since we have

$$L_\gamma = \int_0^1 \frac{|\gamma'(t)|}{1-|\gamma(t)|^2} dt \geq \int_0^1 \frac{\beta'(t)}{1 - \beta(t)^2} dt$$

$$= \int_0^s \frac{du}{1 - u^2} = \frac{1}{2} \log \frac{1 + s}{1 - s} = \text{argtanh} s,$$

it follows that $\rho(0,s) \geqslant \text{argtanh} s$. On the other hand, for $\gamma(t) = ts$, $L_\gamma = \text{argtanh} s$, so that

$$\rho(0,s) = \text{argtanh} s.$$

Since rotations are $\rho$-isometries, we also have

$$\rho(0,z) = \text{argtanh} |z|$$

for all $z \in D$.

If $w \in D$, then $m_{-w}$ is an isometry and $m_{-w}(w) = 0$. Hence

$$\rho(z,w) = \rho(m_{-w}(z), 0) = \text{argtanh} | m_{-w}(z)|.$$

In other words,

$$\rho(z,w) = \text{argtanh} \left|\frac{z - w}{1 - z\bar{w}}\right| .$$

Since $|m_a(z)|^2 = 1 - (1-|a|^2)(1 - |z|^2)/|1 + \bar{a}z|^2$, we can also write

$$\rho(z,w) = \text{argtanh} (1 - \sigma(z,w))^{1/2} ,$$

where

$$\sigma(z,w) = \frac{(1 - |z|^2)(1 - |w|^2)}{|1 - z\bar{w}|^2} .$$

We note that $\rho(u,v) \leqslant \rho(z,w)$ if and only if $\sigma(z,w) \leqslant \sigma(u,v)$, and that $z = w$ if and only if $\sigma(z,w) = 1$.

We are now able to strengthen Theorem 2.1.

<u>Theorem 2.2.</u> If $f \in \mathrm{Hol}(D)$ and there exists a pair of points $z \neq w$ such that $\rho(f(z), f(w)) = \rho(z,w)$, then $f \in \mathrm{Aut}(D)$.

<u>Proof.</u> If $\rho(f(z), f(w)) = \rho(z,w)$, then

$$|m_{-f(w)}(f(z))| = |m_{-w}(z)|,$$

and the result follows from part (d) of the Schwarz-Pick Lemma.

In the next section we begin a more detailed study of the metric space $(D,\rho)$.

## 3. BALLS IN $(D,\rho)$

The metric space $(D,\rho)$ is unbounded. In fact, for any sequence $\{z_n\} \subset D$, $\lim_{n\to\infty} \rho(0,z_n) = \infty$ if and only if $\lim_{n\to\infty} |z_n| = 1$. On the other hand, consider the disc $D_r = \{z \in D : |z| < r\}$, $0 < r < 1$. If $z$ and $w$ belong to $D_r$, then we have

$$\operatorname{argtanh} \frac{|z-w|}{1+r^2} \leq \rho(z,w) = \operatorname{argtanh} |m_{-w}(z)| \leq \operatorname{argtanh} \frac{|z-w|}{1-r^2} .$$

Therefore the topology defined on $D$ by $\rho$ is uniformly equivalent to the standard plane topology on each $D_r$. In particular, the space $(D,\rho)$ is complete and locally compact.

Let $B(a,r)$ be the $\rho$-ball $\{z \in D: \rho(a,z) < r\}$ centered at $a$ and of radius $r$. If $a = 0$, then

$$B(0,r) = \{z \in D : \rho(0,z) < r\} = \{z \in D : |z| < \tanh r\} = D_{\tanh r}.$$

In other words, such balls are discs which are also centered at the origin.

By Theorem 2.1, each $f$ in $\mathrm{Aut}(D)$ is a $\rho$-isometry. Hence

$$B(a,r) = m_a(B(0,r)) = \{z \in D : m_{-a}(z) \in B(0,r)\}$$

$= \{z \in D : |m_{-a}(z)| < \tanh r\}.$

Set $\tanh r = d$. The inequality $\left|\frac{z-a}{1-\bar{a}z}\right| < d$ may be rewritten in the form $|z - b| < r_1$, where

$$b = \frac{1 - d^2}{1 - |a|^2 d^2} a \quad \text{and}$$

$$r_1 = \frac{d(1 - |a|^2)}{1 - |a|^2 d^2}.$$

Thus $B(a,r)$ is also a disc, but it is centered at $b$. Note that the above-mentioned equivalence of the Euclidean and $\rho$ topologies on each $D_r$ shows that closed $\rho$-balls are closed discs and $\rho$-spheres are circles.

For $|u| = 1$ and $0 \leq t < 1$, consider the $\rho$-balls $B(su,\rho(su,tu))$ with $t < s < 1$. When $s \to 1$, we obtain the disc with center $(1+t)u/2$ and radius $(1-t)/2$. The result of the analogous construction in the Euclidean plane leads us to call this disc a half-space (of the first kind). Half-spaces of the second kind will be defined in Section 7.

## 4. METRIC LINES AND $\rho$-CONVEXITY

We say that a mapping $\gamma: R \to D$ is metric embedding of $R$ into $D$ if

$$\rho(\gamma(s), \gamma(t)) = |s-t|$$

for all real $s$ and $t$. The image of $R$ under a metric embedding will be called a metric line. The image of a real interval under such a mapping will be called a metric segment.

Metric embeddings and metric lines do exist. For example, the mapping

$$\gamma: R \to (-1,1).$$

defined by $\gamma(t) = \tanh t$ is a metric embedding of R onto the open interval $(-1,1) \subset D$. Therefore this interval is a metric line. There are, of course, other metric embeddings onto $(-1,1)$. For instance, the mapping $\gamma(t) = m_a(\tanh t)$ is such an embedding whenever $-1 < a < 1$. In general it is clear that the concepts of metric embeddings and metric lines are invariant under ρ-isometries. In other words, if $\gamma$ is a metric embedding, $\Gamma$ is a metric line, and $h: D \to D$ is a ρ-isometry, then $h \circ \gamma$ is also a metric embedding, and $h(\Gamma)$ is a metric line.

It follows, of course, that all sets of the form $h((-1,1))$ with $h \in \mathrm{Aut}(D)$ are metric lines. Such lines are easily described. If $h(0) = 0$, then h is a rotation and

$$h((-1,1)) = e^{i\phi}(-1,1) \text{ for some real } \phi.$$

If $h(0) \neq 0$, then $h((-1,1))$ is an arc of a circle, as an image of a straight segment under a Möbius transformation. Moreover, since any $h \in \mathrm{Aut}(D)$ is conformal on D, the arc $h((-1,1))$ must be orthogonal to the boundary $\partial D$. Note that for any two points z and w in D, there exists a metric line of this form containing them.

As a matter of fact, the above-mentioned arcs are the only metric lines in D. To prove this, it suffices to show that the only metric lines passing through the origin are of the form $e^{i\phi}(-1,1)$. Indeed, let $\Gamma$ be a metric line containing 0, and let a and b belong to $\Gamma$. Assume that $\rho(a,b) = \rho(a,0) + \rho(0,b)$ (the same argument will work if $\rho(b,0) = \rho(b,a) + \rho(a,0)$). Then the closed balls $\overline{B}(a,\rho(a,0))$ and $\overline{B}(b,\rho(b,\phi))$ intersect only at the origin. In other words, we have two closed circles that touch each other at the origin. Their Euclidean centers must therefore be collinear with the origin, and so are their ρ-centers a and b.

We also note that the group Aut(D) acts transitively on the family of metric lines in D. In other words, if $\Gamma_1$ and $\Gamma_2$ are such lines then there exists $h \in \mathrm{Aut}(D)$ such that $h(\Gamma_1) = \Gamma_2$.

Thus we see that any two points z and w in D can be joined by a unique metric segment which is isometric to the interval

$[0,\rho(z,w)]$. We denote this segement by $[z,w]$. For any $0 \leq t \leq 1$ there exists exactly one point $v$ in $[z,w]$ such that $\rho(z,v) = t\rho(z,w)$ and $\rho(v,w) = (1-t)\rho(z,w)$. By analogy with the standard convex combination we shall write $v = (1-t)z \oplus tw$. In particular, $\frac{1}{2}z \oplus \frac{1}{2}w$ will be called the metric center of $[z,w]$.

Metric segments are also the shortest (in the sense of the metric $\rho$) curves joining their endpoints. Therefore we shall also call them geodesic segments. Metric lines will also be called geodesic lines, or simply geodesics.

We shall say that a subset $C \subset D$ is $\rho$-convex if for any two points $z$ and $w$ in $C$, the segment $[z,w]$ is contained in $C$.

The notion of $\rho$-convexity is Aut(D) invariant. That is, if $C$ is $\rho$-convex, then so is $h(C)$ for any $h$ in Aut(D). It is clear that $\rho$-convex sets need not be convex in the Euclidean sense and vice versa. However, the family of $\rho$-convex subsets of $D$ shares the following basic properties with the family of (Euclidean) convex sets:

(a) The intersection of any family of $\rho$-convex sets is $\rho$-convex;

(b) The union of a linearly ordered (by inclusion) family of $\rho$-convex sets is $\rho$-convex;

(c) The $\rho$-closure of $\rho$-convex set is $\rho$-convex.

It is easy to see that the $\rho$-balls centered at the origin are $\rho$-convex. Therefore the above-mentioned invariance under Aut(D) implies that

(d) All balls in $(D,\rho)$ are $\rho$-convex.

It is also clear that the upper half-disc $\{z \in D\colon \operatorname{Im} z > 0\}$ is $\rho$-convex. Hence

(e) Each geodesic divides $D$ into two $\rho$-convex parts.

## 5. UNIFORM CONVEXITY OF $(D,\rho)$

Let $a \in D$, $r > 0$, and $0 < \varepsilon \leq 2$. Consider the ball $B(a,r)$, and let $x$ and $y$ be two points in this ball satisfying $\rho(x,y) \geq \varepsilon r$. Set $u = \frac{1}{2}x \oplus \frac{1}{2}y$. The Möbius transformation $m_{-u}$ maps the geodesic passing through x and y onto a straight line passing through the origin. The points x and y are mapped onto two points symmetric with respect to the origin. Set $b = m_{-u}(a)$ and $w = m_{-u}(x)$. Then we have

$$r \geq \rho(a,x) = \rho(b,w) = \operatorname{argtanh}\,(1 - \sigma(b,w))^{1/2},$$

and

$$r \geq \rho(a,y) = \rho(b,-w) = \operatorname{argtanh}\,(1 - \sigma(b,-w))^{1/2}.$$

Hence

$$\sigma(b,w) = \frac{(1 - |b|^2)(1 - |w|^2)}{|1 - b\bar{w}|^2} \geq 1 - \tanh^2 r,$$

and

$$\sigma(b,-w) = \frac{(1 - |b|^2)(1 - |w|^2)}{|1 + b\bar{w}|^2} \geq 1 - \tanh^2 r.$$

At least one of the denominators of the left-hand sides of these inequalities must exceed $1 + |b|^2|w|^2$. We also have

$$|w| = \tanh \rho(0,w) = \tanh\,(\tfrac{1}{2}\,\rho(x,y)) \geq \tanh(\tfrac{1}{2}\,\varepsilon r).$$

Therefore

$$|b|^2 \leq \frac{\tanh^2 r - \tanh^2(\frac{1}{2}\,\varepsilon r)}{1 - \tanh^2 r\,\tanh^2\,(\frac{1}{2}\,\varepsilon r)} = \tanh(r(1 + \varepsilon/2))\tanh(r(1 - \varepsilon/2)),$$

so that

$$\rho(a, \tfrac{1}{2}\,x \oplus \tfrac{1}{2}\,y) = \rho(0,b)$$

$$\leq \operatorname{argtanh} \sqrt{\tanh(r(1 + \varepsilon/2))\,\tanh(r(1-\varepsilon/2))}.$$

We now define a function $\delta\colon (0,\infty) \times [0,2] \to (0,1]$ by

$$\delta(r,\varepsilon) = 1 - \frac{1}{r} \text{ argtanh } \sqrt{\tanh(r(1+\varepsilon/2))\tanh(r(1-\varepsilon/2))}$$

to obtain

$$\rho(a, \frac{1}{2} x \oplus \frac{1}{2} y) \leqslant (1 - \delta(r,\varepsilon))r .$$

All these inequalities are sharp. That is, the points a,x, and y can be chosen in such a way that they become equalities. Therefore $\delta$ may also be defined by

$$\delta(r,\varepsilon) = \inf\{1 - \frac{1}{r}\, \rho(a, \frac{1}{2} x \oplus \frac{1}{2} y)\} ,$$

where the infimum is taken over all points a,x and y satisfying $\rho(a,x) \leqslant r$, $\rho(a,y) \leqslant r$, and $\rho(x,y) \geqslant \varepsilon r$.

The function $\delta(r,\varepsilon)$ is analogous to the modulus of convexity in (uniformly convex) Banach spaces (which, however, is independent of r). Therefore we shall call it the modulus of convexity of $(D,\rho)$. It has the following properties:

(a) $\delta(r,0) = 0$ and $\delta(r,2) = 1$ for all $r > 0$;

(b) $\lim_{r\to 0} \delta(r,\varepsilon) = 1 - (1 - \varepsilon^2/4)^{1/2} = \delta_H(\varepsilon)$, the modulus of convexity of Hilbert space;

(c) $\lim_{r\to\infty} \delta(r,\varepsilon) = 0$;

(d) $\delta(r,\varepsilon)$ increases with $\varepsilon$ (for a fixed r), and decreases with respect to r (for a fixed $\varepsilon$).

In addition, we have of course the following implication:

$$\left.\begin{array}{l} \rho(a,x) \leqslant r \\ \rho(a,y) \leqslant r \\ \rho(x,y) \geqslant \varepsilon r \end{array}\right\} \Rightarrow \rho(a, \frac{1}{2} x \oplus \frac{1}{2} y) \leqslant (1 - \delta(r,\varepsilon))r.$$

We conclude that all the balls in $(D,\rho)$ are uniformly rotund. Their rotundity depends, however, on their radii. Balls with equal radii are equally rotund. The rotundity of small balls is more or less the same as the rotundity of balls in Hilbert space, while large balls are almost square.

## 6. NONEXPANSIVE MAPPINGS IN $(D,\rho)$

In this section we intend to take a closer look at the family $N = N(D)$ of $\rho$-nonexpansive self-mappings of $D$. Thus $f:D \to D$ belongs to $N$ if

$$\rho(f(z),f(w)) \leq \rho(z,w)$$

for all $z$ and $w$ in $D$. For $0 \leq k < 1$, we shall also consider the subset $N_k$ of $N$ consisting of all strict $\rho$-contractions with constant $k$. That is, $f:D \to D$ belongs to $N_k$ if

$$\rho(f(z),f(w)) \leq k\rho(z,w)$$

for all $z$ and $w$ in $D$. It is clear that $\mathrm{Hol}(D) \subset N(D)$. As a matter of fact, Theorem 2.2 shows that if $f \in \mathrm{Hol}(D)$ is not an automorphism, then

$$\rho(f(z),f(w)) < \rho(z,w)$$

for all $z \neq w$ in $D$. In order to proceed we need the following fact.

<u>Lemma 6.1.</u> For any $0 \leq k < 1$ and $z,w \in D$,

$$\rho(kz,kw) \leq k\rho(z,w).$$

<u>Proof.</u> Let $\gamma : [0,1] \to D$ be a piecewise continuously differentiable curve joining $z$ and $w$. We have

$$\frac{|k\gamma'(t)|}{1 - k^2|\gamma(t)|^2} = k \frac{1 - |\gamma(t)|^2}{1 - k^2|\gamma(t)|^2} \frac{|\gamma'(t)|}{1 - |\gamma(t)|^2} \leqslant k \frac{|\gamma'(t)|}{1 - |\gamma(t)|^2}.$$

Hence

$$L_{k\gamma} \leqslant kL_\gamma$$

and the result follows.

Choosing $\gamma$ to be the geodesic joining z and w, we obtain even a stronger inequality for $z \neq w$: $\rho(kz,kw) < k\rho(z,w)$.

We conclude that all mappings of the form $f(z) = h^{-1}(kh(z))$ with h in Aut(D) belong to $N_k$, $0 \leqslant k < 1$.

We shall say that $f:D \to D$ maps D strictly inside D if

$$\sup\{|f(z)| : z \in D\} < 1.$$

The following result shows that any $f \in \text{Hol}(D)$ that maps D strictly inside D is a strict $\rho$-contraction.

Theorem 6.2. If $f \in \text{Hol}(D)$ and $k = \sup\{|f(z)|: z \in D\} < 1$, then $f \in N_k$ with $\rho(f(z),f(w)) < k\rho(z,w)$ for $z \neq w$.

Proof. The result is clear if f is constant. If it is not constant, then the maximum principle shows that f/k belongs to Hol(D). Hence

$$\rho(f(z),f(w)) = \rho(kf(z)/k,\ kf(w)/k) \leqslant k\rho(z,w),$$

with strict inequality if $z \neq w$.

Lemma 6.3. If the points z,w,u and v belong to D, and $\rho(u,v) \leqslant \rho(z,w)$, then there exists a mapping f in Hol(D) such that $f(z) = u$ and $f(w) = v$.

Proof. There are mappings g and h in Aut(D) such that $g(z) = 0$, $g(w) = r_1 \geq 0$, $h(u) = 0$ and $h(v) = r_2 \geq 0$. Since $r_2 \leq r_1$, the required f may be defined by $f(z) = h^{-1}((r_2/r_1)g(z))$ for z in D.

Lemma 6.4. If the points z,w,u and v belong to D, then $\rho((1-t)z + tu, (1-t)w + tv) \leq \max (\rho(z,w), \rho(u,v))$ for all $0 \leq t \leq 1$.

Proof. Denote $\max(\rho(z,w), \rho(u,v))$ by a, and choose points x and y in D such that $\rho(x,y) = a$. By Lemma 6.3, there are mappings f and g in Hol(D) such that $f(x) = z$, $f(y) = w$, $g(x) = u$, and $g(y) = v$. then $h = (1-t)f + tg$ belongs to Hol(D), and

$$\rho((1-t)z + tu, (1-t)w + tv) = \rho(h(x), h(y)) \leq \rho(x,y) = a.$$

We conclude that the family N is convex:

Theorem 6.5. If f and g belong to N, then so does $(1-t)f + tg$ for each $0 \leq t \leq 1$.

As a matter of fact, N is convex not only in the Euclidean sense, but also in a metric sense. We shall establish this claim by proving a series of lemmas.

Lemma 6.6. If x,y and z are not on the same metric line, then

$$\rho(\tfrac{1}{2} x \oplus \tfrac{1}{2} z, \tfrac{1}{2} y \oplus \tfrac{1}{2} z) < \tfrac{1}{2} \rho(x,y) .$$

Proof. We may and shall assume that $z = 0$. Note that if $v = \frac{1}{2} w \oplus \frac{1}{2} 0$, then $w = 2v/(1 + |v|^2)$. Now let $\gamma: R \to D$ be a metric embedding with $\gamma(0) = x$ and $\gamma(1) = y$. Set

$$x_1 = \frac{1}{2} x \oplus \frac{1}{2} 0 \ , \ y_1 = \frac{1}{2} y \oplus \frac{1}{2} 0,$$

and consider the curve $\gamma_1$ defined by $\gamma_1(t) = \frac{1}{2} \gamma(t) \oplus \frac{1}{2} 0$. Then

$$\gamma_1(0) = x_1, \ \gamma_1(1) = y_1 \ , \text{ and } \gamma(t) = 2\gamma_1(t)/(1 + |\gamma_1(t)|^2).$$

Note that $\gamma_1$ is not a geodesic. Now we have

$$\rho(x,y) = L_\gamma = \int_0^1 \frac{|\gamma'(t)|}{1 - |\gamma(t)|^2} \, dt$$

$$= 2 \int_0^1 \frac{|\gamma_1'(1 + |\gamma_1|^2) - 2\gamma_1 \operatorname{Re} \overline{\gamma}_1 \gamma_1'|}{(1 - |\gamma_1|^2)^2} \, dt$$

$$\geq 2\int_0^1 \frac{|\gamma_1'|}{1 - |\gamma_1|^2} \, dt = 2L_{\gamma_1} > 2\rho(x_1, y_1),$$

as claimed.

<u>Lemma 6.7.</u> If x,y,z and w are not on the same metric line, then

$$\rho\left( \frac{1}{2} x \oplus \frac{1}{2} z \ , \ \frac{1}{2} y \oplus \frac{1}{2} w\right) < \frac{1}{2} \rho(x,y) + \frac{1}{2} \rho(z,w).$$

Proof. We have

$$\rho\left( \frac{1}{2} x \oplus \frac{1}{2} z, \ \frac{1}{2} y \oplus \frac{1}{2} w\right)$$

$$\leq \rho\left(\frac{1}{2} x \oplus \frac{1}{2} z, \ \frac{1}{2} y \oplus \frac{1}{2} z\right)$$

$$+ \rho\left(\frac{1}{2} y \oplus \frac{1}{2} z, \ \frac{1}{2} y \oplus \frac{1}{2} w\right) < \frac{1}{2} \rho(x,y) + \frac{1}{2} \rho(z,w)$$

by Lemma 6.6.

Lemma 6.8. If $x,y,z$ and $w$ are in $D$ and $0 \leq t \leq 1$, then

$$\rho((1-t)x \oplus tz,\ (1-t)y \oplus tz) \leq (1-t)\ \rho(x,y),$$

and

$$\rho((1-t)x \oplus tz,\ (1-t)y \oplus tw) \leq (1-t)\rho(x,y) + t\rho(z,w).$$

Proof. We first use Lemmas 6.6 and 6.7, as well as induction, to prove this result for $t = k/2^n$, where $n$ and $k$ are integers. We then complete the proof by using a continuity argument.

Thus we obtain the following metric convexity result for N.

Theorem 6.9. If $f \in N_k$, $g \in N_\ell$, and $0 \leq t \leq 1$, then $(1-t)f \oplus tg$ belongs to $N_{(1-t)k+t\ell}$. In particular, if $f$ and $g$ belong to N, then so does $(1-t)f \oplus tg$.

The family N contains many mappings which do not belong to Hol(D). since the mapping $z \to \bar{z}$ is a non-holomorphic $\rho$-isometry of D onto itself, all anti-holomorphic mappings, that is mappings of the form $f(\bar{z})$ with $f \in \mathrm{Hol}(D)$, belong to N. By Theorems 6.5 and 6.9, N also contains all convex and metric convex combinations of holomorphic and anti-holomorphic mappings. Among these combinations are two $\rho$-nonexpansive mappings of D onto $(-1,1) \subset D$ which keep all the points of $(-1,1)$ fixed. They are $\mathrm{Re}z = (z + \bar{z})/2$ and

$$P(z) = \frac{1}{2} z \oplus \frac{1}{2} \bar{z}\ .$$

These mappings are neither holomorphic nor anti-holomorphic. For each $0 \leq t \leq 1$, the mappings $(1-t)P(z) + t\mathrm{Re}z$ and $(1-t)P(z) \oplus t\mathrm{Re}z$ are also $\rho$-nonexpansive retractions of D onto the geodesic $(-1, 1)$. Thus there are infinitely many such retractions. This is in sharp contrast with the situation in Hilbert space, where there is only one nonexpansive retraction onto a straight line.

We know that the nearest point projection onto any closed convex subset of a Hilbert space is nonexpansive. We now show that this fact has an analog in $(D,\rho)$.

Let $C$ be a $\rho$-closed $\rho$-convex subset of $D$. For any $z \in D$, let

$$\mathrm{dist}(z,C) = \inf\{\rho(z,u) : u \in C\}.$$

Local compactness shows that there exists a point $v$ in $C$ such that $\rho(z,v) = \mathrm{dist}(z,C)$. The $\rho$-convexity of $C$ and the uniform convexity of the balls in $D$ show that $v$ is unique. Thus we may define the nearest point projection $P_C\colon D \to C$ by $P_C(z) = v$. If $C = (-1,1)$, then $P_C$ assigns to each point $z \in D$ the point $t \in (-1,1)$ satisfying

$$\frac{t}{1+t^2} = \frac{\mathrm{Re}z}{1+|z|^2} .$$

In this case $P_C(z)$ coincides with the metric center of $z$ and $\bar{z}$, and we have $P_{(-1,1)}(z) = P_{(-1,1)}(\bar{z}) = \frac{1}{2} z \oplus \frac{1}{2} \bar{z} = P(z)$. If $\gamma$ is any other geodesic in $D$, then $P_\gamma = h^{-1} \circ P \circ h$ where $h$ is any automorphism mapping $\gamma$ onto $(-1,1)$. Hence each $P_\gamma$ belongs to $N$. The general result is

Theorem 6.10. The nearest point projection onto a $\rho$-closed $\rho$-convex subset of $(D,\rho)$ is $\rho$-nonexpansive.

Proof. Let $C$ be a $\rho$-closed $\rho$-convex subset of $(D,\rho)$, and let $P_C\colon D \to C$ be the nearest point projection onto $C$. Let $x$ and $y$ be two points in $D$, and let $\gamma$ be the geodesic joining $P_C(x)$ and $P_C(y)$. Note that the metric segment joining $P_C(x)$ and $P_C(y)$ must be contained in the metric segment joining $P_\gamma(x)$ and $P_\gamma(y)$. Hence

$$\rho(P_C(x),P_C(y)) \le \rho(P_\gamma(x),\ P_\gamma(y)) \le \rho(x,y).$$

Ascoli's theorem shows that every sequence $\{f_n\} \subset N$ contains a subsequence which is pointwise convergent in the Euclidean topology of D. (As a matter of fact, the convergence is uniform on each compact subset of D.) However, the limit of such a sequence does not necessarily map D into D. If it does, then it also belongs to N.

The last result of this section shows that N does not contain any "metric similarities".

Proposition 6.11. If $f:D \to D$ and $\rho(f(x),f(y)) = k\rho(x,y)$ for some $0 \leq k \leq 1$ and all x and y in D, then either $k = 0$ or $k = 1$.

Proof. Assuming the $k > 0$ we observe that such a mapping f transforms metric lines onto metric lines. As a matter of fact, $f((1-t)x \oplus ty) = (1-t)f(x) \oplus tf(y)$ for all x and y in D and $0 \leq t \leq 1$. We may assume that $f(0) = 0$ and that $f((-1,1)) = (-1,1)$. Since $f(P_C(x)) = P_{f(C)}(f(x))$ for all nearest point projections $P_C$, the imaginary axis is also invariant under f. When we compare now the distances among the points $a, ai, f(a)$ and $f(ai)$, $-1 < a < 1$, we see that k must equal 1.

## 7. EQUIDISTANT SETS AND HALF-SPACES

If x and y are two distinct points in D, we define the equidistant set $Eq(x,y)$ by

$$Eq(x,y) = \{z \in D: \ \rho(z,y) = \rho(z,x)\}.$$

We also define two closed half-spaces (of the second kind) by

$$Eq^+(x,y) = \{z \in D: \ \rho(z,y) \leq \rho(z,x)\}$$

and

$$E\bar{q}(x,y) = E\overset{+}{q}(y,x).$$

If $y = -x$, then

$$Eq(x,y) = \{z \in D\colon \operatorname{Re} z\bar{x} = 0\}$$

and

$$E\overset{+}{q}(x,y) = \{z \in D\colon \operatorname{Re} z\bar{x} < 0\}.$$

Thus $Eq(x,y)$ is a metric line passing through the origin which is perpendicular to x. It follows that any equidistant set is a metric line. For any two points x and y in D,

$$Eq(x,y) = \{z \in D\colon \operatorname{Re} m_{-u}(z)\overline{m_{-u}(x)} = 0\},$$

where $u = \frac{1}{2}x \oplus \frac{1}{2}y$. This metric line passes through u and is orthogonal to the metric segment $[x,y]$. It is also clear that

$$E\overset{+}{q}(x,y) = \{z \in D\colon \operatorname{Re} m_{-u}(z)\overline{m_{-u}(x)} < 0\}.$$

In preparation for subsequent results we also wish to consider the set $A(x,y)$ consisting of all $z \in D$ for which the function $\phi_z : [0,1] \to R^+$ defined by

$$\phi_z(t) = \rho(z,(1-t)x \oplus ty)$$

is non-increasing. Since $\phi_z$ is convex, this requirement is equivalent to $\phi_z'(1) \leq 0$. In other words, $z \in A(x,y)$ if $\tilde{\phi}_z(t) = \sigma(z,(1-t)x \oplus ty)$ is non-decreasing for $0 \leq t \leq 1$, or, equivalently, if $\tilde{\phi}_z'(1) \geq 0$. Let $2y \ominus x$ denote the unique point u on the metric line passing through x and y which satisfies

$$y = \frac{1}{2}x \oplus \frac{1}{2}u.$$

(Thus $2y \ominus x$ is the point metrically symmetric to x with respect to y.) If $w = m_{-y}(z)$ and $v = m_{-y}(x)$, then $\tilde{\phi}_z(t) = \sigma(w, s(t)v)$, where $s:[0,1] \to [0,1]$ is a decreasing function of t with $s(0) = 1$ and $s(1) = 0$. Hence $\tilde{\phi}_z'(1) \geq 0$ is equivalent to

$$\frac{d}{dp} \sigma(w, pv)\Big|_{p=0} \leq 0.$$

Since this derivative equals $2(1 - |w|^2)\mathrm{Re}\, w\bar{v}$, we see that $z \in A(x,y)$ if and only if $w \in \overset{+}{\mathrm{Eq}}(v,-v)$. Therefore

$$A(x,y) = m_y(\overset{+}{\mathrm{Eq}}(v,-v)) = \overset{+}{\mathrm{Eq}}(x, 2y \ominus x).$$

Consider now the set $B(x,y)$ consisting of all points $z \in D$ for which the function $\psi_z: [0,1] \to R^+$ defined by

$$\psi_z(t) = \rho(z, (1-t)x + ty)$$

is non-increasing. Equivalently, $z \in B(x,y)$ if

$$\tilde{\psi}_z(t) = \sigma(z, (1-t)x + ty))$$

is non-decreasing for $0 \leq t \leq 1$. Therefore a necessary condition for this to happen is that $\psi_z'(1) \geq 0$. A computation shows that this latter condition is equivalent to

$$\mathrm{Re}\; m_{-y}(z)\overline{(x-y)} \leq 0, \tag{7.1}$$

or in other words,

$$\mathrm{Re}\; m_{-y}(z)\; \overline{m_{-y}(x)}\; (1 - y\bar{x}) \leq 0. \tag{7.2}$$

The set of all $z \in D$ that satisfy (7.1) (or (7.2)) is a closed half-space of the second kind which is determined by the metric line passing through y that is orthogonal to the Euclidean segment I joining x and y. It can also be obtained by considering the sets $A((1-t)x + ty,y)$ and letting $t \to 1-$. Consider now any point $u \neq y$ on I. It is not difficult to see that the metric line passing through u that is orthogonal to I cannot intersect the corresponding line through y. It follows that (7.1) is not only a necessary condition for z to belong to B(x,y), but also a sufficient one. In other words,

$$B(x,y) = \{z \in D: \operatorname{Re} m_{-y}(z)\overline{(x-y)} \leqslant 0\}.$$

Note that if $z \in B(x,y)$ and $z \neq y$, then $|z-x| > |y - x|$.

Intuitively speaking, the metric segment joining x to y is "directed" toward the half-space A(x,y), while the Euclidean segment joining these two points is "directed" toward B(x,y). This is analogous to the Euclidean situation. If x and y are two distinct complex numbers, then the function $|z - (1-t)x - ty|$ decreases for $0 \leqslant t \leqslant 1$ if and only if z belongs to the closed half-plane determined by the line passing through y and perpendicular to $y - x$, in other words, if and only if

$$\operatorname{Re} (z-y)\overline{(x-y)} \leqslant 0.$$

We conclude this section with another representation of the half-spaces $\overset{+}{Eq}(x,y)$. The inequality $\rho(z,y) \leqslant \rho(z,x)$ is equivalent to

$$\left|\frac{1 - y\bar{z}}{1 - x\bar{z}}\right| \leqslant \left(\frac{1 - |y|^2}{1 - |x|^2}\right)^{1/2} = c,$$

which implies that

$$\frac{1 - y\bar{z}}{1 - x\bar{z}} = \xi , \qquad (7.3)$$

where $\xi$ is a complex number such that $|\xi| \leqslant c$.

We rewrite (7.3) as

$$(x + \frac{d}{1-\xi} v)\bar{z} = 1, \tag{7.4}$$

where $d = |y - x|$ and $v = (y - x)/|y-x|$. The mapping

$$\xi \to \eta = x + \frac{d}{1 - \xi} v$$

maps the disc $|\xi| < c$ onto a disc, a half-plane, or the complement of a disc if $c < 1$, $c = 1$, or $c > 1$ respectively. In each case the image is invariant under inversion with respect to D. $E_q^+(x,y)$ is the intersection of this image with D.

## 8. ISOMETRIC MODELS OF $(D,\rho)$

The space $(D,\rho)$ has many interesting isometric images. By the Riemann Mapping Theorem, every simply connected domain U in the plane (except the plane itself) is conformally equivalent to the open unit disc D. This means that there exists a holomorphic mapping $g : D \to U$ such that g is one-to-one and $g(D) = U$. Therefore we can define a metric $\rho_U$ on U by

$$\rho_U(x,y) = \rho(g^{-1}(x), g^{-1}(y)).$$

The space $(U,\rho_U)$ is isometric to $(D,\rho)$ and g is an isometry. We shall call $\rho_U$ the Poincaré metric on U. Clearly the automorphisms of $(U,\rho_U)$ are of the form $g \circ h \circ g^{-1}$ with $h \in \mathrm{Aut}(D)$, the geodesics are of the form $g \circ \gamma$ with $\gamma$ a geodesic in D, and the modulus of convexity $\delta_U = \delta$. We shall now briefly discuss two particular models.

Let $\Pi^+$ denote the upper half-plane $\{z \in C: \mathrm{Im}\, z > 0\}$. The function

$$g(z) = i\,\frac{1 + z}{1 - z}$$

is one-to-one and maps D onto $\Pi^+$. It is also well-defined on $\partial D \setminus \{1\}$ and maps this arc onto the real line. The point 1 is sent to infinity. The inverse function is

$$g^{-1}(w) = \frac{w - i}{w + i}.$$

Hence

$$\rho_{\Pi^+}(x,y) = \operatorname{argtanh} \left|\frac{x-y}{\bar{x}-y}\right|.$$

Balls in $(\Pi^+, \rho_{\Pi^+})$ are again discs. Metric lines are either semi-circles with their end-points on the real axis, or half-lines of the form Rez = constant. These half-lines are the images under g of those metric lines in D with 1 as one of their "points at infinity". Since the automorphisms of D have continuous extensions to $\bar{D}$, the automorphisms of $\Pi^+$ have continuous extensions to the closed extended upper half-plane $\overline{\Pi^+} \cup \{\infty\}$. The image of the point $\{\infty\}$ by such an automorphism is either the point $\{\infty\}$ itself or a point on the real axis. The upper half-plane model can be used to gain a better understanding of $(D,\rho)$. For example, consider a line of the form Im z = constant in $\Pi^+$. Such a line is orthogonal to all the metric lines "containing" $\{\infty\}$. Thus the image of this line under $g^{-1}$ is orthogonal to all the metric lines in D "containing" 1. On the other hand this image must be a circle containing only one point of $\partial D$. It follows that the orthogonal trajectories of the family of all geodesics having a common "point at infinity" are circles tangent to $\partial D$.

Consider the Koebe function $K : D \to C \setminus (-\infty, -\frac{1}{4})$ defined by $K(z) = z/(1-z)^2$. It has a continuous extension to $\partial D \setminus \{1\}$ and $K(1) = \infty$. The Koebe model $U = K(D)$ is of interest because it is not convex. Thus the family $N_U$ of $\rho_U$-nonexpansive mappings is no longer convex (in the Euclidean sense), and $\rho_U$-balls are not necessarily convex sets. However the space $(U,\rho_U)$ is again $\rho_U$-uniformly convex with the same modulus of convexity $\delta_U = \delta$ as the unit disc.

## 9. HOLOMORPHIC MAPPINGS IN BANACH SPACES

Our aim in this section is to briefly discuss the basic properties of holomorphic mappings acting between Banach spaces. More details can be found, for example, in [36] and [52]. All the Banach spaces involved in this discussion are assumed to be complex.

By a domain we shall mean a nonempty connected open subset of a Banach space. A function f mapping a domain $\mathcal{D}$ in a Banach space X into a Banach space Y is said to be holomorphic if it is Fréchet differentiable at each x in $\mathcal{D}$. The Fréchet derivative Df(x) of f at x is a bounded (complex) linear operator of X into Y. We have of course

$$Df(x)v = \lim_{\xi \to 0} \frac{f(x+\xi v) - f(x)}{\xi}$$

for each v in X.

It is clear that any linear mapping is holomorphic and that the composition of two holomorphic mappings is also holomorphic. If f is holomorphic, then all higher order Fréchet derivatives $D^n f(x)$ also exist. For each n, they are symmetric multilinear mappings from $X^n$ into Y. With each derivative we may associate a "homogeneous polynomial" defined by

$$\hat{D}^n f(x)v = D^n f(x)(v,v,\dots,v)$$

for each v in X. Let $B(x_0,r)$ denote the open ball in X with center $x_0$ and radius r. We quote now Taylor's Theorem.

<u>Theorem 9.1.</u> If f: $B(x_0,r) \to Y$ is a bounded holomorphic function, then

$$f(x) = \sum_{n=0}^{\infty} \frac{1}{n!} \hat{D}^n f(x_0)(x-x_0)$$

for all x in $B(x_0,r)$. The convergence of the above series is uniform on $B(x_0,s)$ for each $s < r$.

Next we prove a generalization of the Schwarz Lemma (Theorem 1.2). Norms will be denoted by $|\cdot|$.

Theorem 9.2. If $f:B(0,r) \to Y$ is a holomorphic function such that $|f(x)| \leq M$ for all $x \in B(0,r)$ and $f(0) = 0$, then $|f(x)| \leq M|x|/r$ for all $x \in B(0,r)$.

Proof. Let $x \neq 0$ belong to $B(0,r)$, and let $\phi$ be a linear functional on Y of norm 1 such that $|\phi(f(x))| = |f(x)|$. Define $g: D \to D$ by

$$g(\xi) = \phi(f(\xi rx/|x|))/M .$$

Since $g \in \mathrm{Hol}(D)$ and $g(0) = 0$, the one-dimensional Schwarz Lemma shows that $|g(\xi)| \leq |\xi|$ for all $\xi \in D$. We now complete the proof by choosing $\xi = |x|/r$.

Theorem 9.2 shows, in particular, that if f maps the unit ball of X into the closed unit ball of Y and $f(0) = 0$, then $|f(x)| \leq |x|$ for all x in $B(0,1)$.

It also shows that a bounded holomorphic mapping defined on the whole space X must be constant. This is of course a generalization of Liouville's theorem on entire functions.

We also quote the following extensions of the Cauchy estimates and the identity theorem.

Theorem 9.3. If $f:B(x_0,r) \to Y$ is a holomorphic function such that $|f(x)| \leq M$ for all x in $B(x_0,r)$, then

$$\frac{1}{n!}\,|\hat{D}^n f(x_0)v| \leq \frac{M|v|^n}{r^n}$$

for all $n = 1,2,\ldots$ and $v \in X$.

Theorem 9.4. If $f:\mathcal{D} \to Y$ is a holomorphic function vanishing identically on an open ball in $\mathcal{D}$, then it is identically zero on all of $\mathcal{D}$.

Thus we see that the properties of holomorphic functions between Banach spaces are similar to those of holomorphic functions in the complex plane. There are, however, important differences. The maximum principle, for example, does not carry over. To see this, let $X = C$ and $Y = C^2$ be equipped with the norm

$$|(y_1,y_2)| = \max\,\{|y_1|\;,\;|y_2|\}.$$

The function $f\colon D \to \overline{B}(0,1) = \overline{D} \times \overline{D} \subset Y$ defined by $f(x) = (1,x)$ is holomorphic with $|f(x)| = 1$ for all $x \in D$, but it is not constant. If Y is a "nicer" space (for instance, if it is strictly convex), then the maximum principle can be extended. See [36] and [94].

## 10. THE CARATHÉODORY-REIFFEN-FINSLER PSEUDOMETRIC

In this section we extend the construction that led to the definition of the Poincaré metric on the open unit disc D to domains in Banach spaces.

For such a domain $\mathcal{D} \subset X$ let $\mathrm{Hol}(\mathcal{D},D)$ be the family of all holomorphic functions $f\colon \mathcal{D} \to D$. The Cauchy estimates show that for any such f and any $x_0$ in $\mathcal{D}$,

$$|Df(x_0)v| \leq \frac{|v|}{\mathrm{dist}(x_0,\partial\mathcal{D})}.$$

Therefore we may define a locally bounded lower semicontinuous function $\alpha : \mathcal{D} \times X \to R$ by

$$\alpha(x,v) = \sup\{|Dg(x)v| : g \in \mathrm{Hol}(\mathcal{D}, D)\}.$$

This function is called the infinitesimal Carathéodory-Reiffen-Finsler pseudometric on $\mathcal{D}$. For $\mathcal{D} = D$, we have

$$\alpha(x,v) = |v|/(1 - |x|^2).$$

To distinguish between the pseudometrics assigned to different domains $\mathcal{D}_1, \mathcal{D}_2, \ldots$, we shall use the notations $\alpha_{\mathcal{D}_1}, \alpha_{\mathcal{D}_2}, \ldots$, and $\alpha_1, \alpha_2, \ldots$

<u>Lemma 10.1.</u> If $f\colon \mathcal{D}_1 \to \mathcal{D}_2$ is a holomorphic function between the domains $\mathcal{D}_1 \subset X_1$ and $\mathcal{D}_2 \subset X_2$, then

$$\alpha_2(f(x), Df(x)v) \le \alpha_1(x,v)$$

for all $x \in \mathcal{D}_1$ and $v \in X_1$.

<u>Proof.</u> If $g \in \mathrm{Hol}(\mathcal{D}_2, D)$, then $h = g \circ f$ belongs to $\mathrm{Hol}(\mathcal{D}_1, D)$. Therefore the chain rule shows that $|Dg(f(x))Df(x)v| = |Dh(x)v| \le \alpha_1(x,v)$. The result follows.

This lemma is an extension of part (b) of Theorem 1.4.

Given any two points x and y in $\mathcal{D}$, consider the (nonempty) family of all curves $\gamma\colon [0,1] \to \mathcal{D}$ that join x and y and have piecewise continuous derivatives. Call such a curve admissible and define its "length" by

$$L(\gamma) = \int_0^1 \alpha(\gamma(t), \gamma'(t))dt .$$

We now define the pseudo-distance between x and y by $\rho(x,y) = \inf\{L(\gamma)\colon \gamma \text{ is admissible with } \gamma(0) = x \text{ and } \gamma(1) = y\}$.

It is easy to prove that $\rho$ is in fact a pseudometric on $\mathcal{D}$. We shall call it the CRF pseudometric. In general $\rho$ is not a metric. To see this, let $\mathcal{D} = C$. Any holomorphic (entire) function $f: C \to D$ is constant. Hence $\alpha_C \equiv 0$ and so is $\rho_C$. If $\mathcal{D} = D$, then $\rho$ is the Poincaré metric. We now establish the main property of the CRF pseudometrics.

Theorem 10.2. If $f: \mathcal{D}_1 \to \mathcal{D}_2$ is a holomorphic function, then $\rho_2(f(x),f(y)) \leq \rho_1(x,y)$ for all x and y in $\mathcal{D}_1$.

Proof. If $\gamma$ is an admissible curve joining x and y, then $\beta = f \circ \gamma$ is an admissible curve joining f(x) and f(y). Since $\alpha_2(\beta(t),\beta'(t)) = \alpha_2(f(\gamma(t)), Df(\gamma(t))\gamma'(t)) \leq \alpha_1(\gamma(t), \gamma'(t))$ by Lemma 10.1, $L(\beta) \leq L(\gamma)$ and the result follows.

We now use Theorem 10.2 to prove our next result.

Theorem 10.3. If $\mathcal{D} \subset X$ is a bounded domain, then $\rho$ is a metric.

Proof. It is enough to show that if $x \neq y$ then $\rho(x,y) > 0$. To this end, let $\phi$ be a bounded linear functional on X of norm 1 such that $\phi(x-y) = |x-y|$. Let d be the diameter of $\mathcal{D}$, and define

$$g: \mathcal{D} \to D \text{ by } g(x) = \phi((x-y)/d).$$

Since g is holomorphic, Theorem 10.2 shows that

$$\rho(x,y) \geq \rho_D(g(x),g(y)) = \rho_D(g(x), 0)$$

$$= \operatorname{argtanh}(|x-y|/d) > 0.$$

The inequality $|x-y| \leq d \tanh(\rho(x,y))$ obtained in the proof of Theorem 10.3 shows that the norm metric is weaker than the CRF metric. In the other direction, we have

Theorem 10.4. If $\mathcal{D} \subset X$ is a bounded domain, then

$$\rho(x,y) \leq \operatorname{argtanh}(|x-y|/\operatorname{dist}(x,\partial\mathcal{D}))$$

whenever $|x-y| < \operatorname{dist}(x, \partial\mathcal{D})$.

Proof. Fix $x \in \mathcal{D}$ and let $y \in \mathcal{D}$ satisfy $|x-y| < \operatorname{dist}(x,\partial\mathcal{D})$. Since the function $f: D \to \mathcal{D}$ defined by

$$f(\xi) = x + \xi(y-x)\operatorname{dist}(x,\partial\mathcal{D})/|y-x|$$

is holomorphic,

$$\rho_{\mathcal{D}}(x,y) = \rho_{\mathcal{D}}(f(0), f(|y-x|/\operatorname{dist}(x,\partial\mathcal{D})))$$

$$\leq \rho_D(0,|x-y|/\operatorname{dist}(x, \partial\mathcal{D}))$$

$$= \operatorname{argtanh}(|x-y|/\operatorname{dist}(x,\partial\mathcal{D})),$$

and the proof is complete.

Theorems 10.3 and 10.4 show that on a bounded domain $\mathcal{D}$ the CRF metric $\rho$ is topologically equivalent to the norm metric. This equivalence is even locally uniform. But it is not enough to ensure completeness of $\rho$. Any Cauchy sequence with respect to $\rho$ is also Cauchy with respect to the norm, but its norm limit may belong to the boundary of $\mathcal{D}$. Note that for such a sequence $\{x_n\}$, $\lim_{n\to\infty} x_n = x \in \partial\mathcal{D}$ (in the norm topology), but $\limsup_{n\to\infty} \rho(x_n,y) < \infty$ for all $y \in \mathcal{D}$. Thus not all points on the boundary of $\mathcal{D}$ are at infinity. Examples of these phenomena may be found in [48].

## 11. SCHWARZ-PICK SYSTEMS OF PSEUDOMETRICS

By defining the CRF pseudometrics we have assigned to each domain in every Banach space a pseudometric. This system of pseudometrics has the following two properties:

The pseudometric assigned to D is the Poincaré metric; (11.1)

If $\rho_1$ and $\rho_2$ are assigned to $\mathcal{D}_1$ and $\mathcal{D}_2$ respectively, then $\rho_2(f(x),f(y)) \leq \rho_1(x,y)$ for all holomorphic $f: \mathcal{D}_1 \to \mathcal{D}_2$ and all x and y in $\mathcal{D}_1$. (11.2)

We shall call any system of pseudometrics that satisfies these two conditions a Schwarz-Pick system [48]. The proofs of Theorems 10.3 and 10.4 show that if $\rho$ is assigned to $\mathcal{D}$ by a y Schwarz-Pick system, and $\mathcal{D}$ is contained in the open ball B(x,R), then

$$\rho(x,y) \geq \operatorname{argtanh}(|x-y|/R)$$

for all y in $\mathcal{D}$. Also, if $B(x,r) \subset \mathcal{D}$ and $y \in B(x,r)$ then

$$\rho(x,y) \leq \operatorname{argtanh}(|x-y|/r).$$

Consequently, if $\mathcal{D} = B(x,r)$, then $\rho(x,y) = \operatorname{argtanh}(|x-y|/r)$ for all y in $\mathcal{D}$. In particular,

$$\rho(0,y) = \operatorname{argtanh} |y|$$

when $\mathcal{D} = B(0,1)$, the unit ball of X. This agrees, of course, with the one-dimensional case. We draw two immediate conclusions from these facts.

The first is that any open ball in a Banach space is complete with respect to any metric assigned to it by a Schwarz-Pick system. The second is that the intersection of a ball with a straight line passing through its center is a metric line for such a metric.

There are other Schwarz-Pick systems in addition to the CRF pseudometrics. We now describe two such systems. Let $\mathcal{D}$ be a domain in a Banach space. For any $f \in Hol(\mathcal{D},D)$ and $x,y \in \mathcal{D}$

$$\rho_D(f(x),f(y)) \leq \rho(x,y), \tag{11.3}$$

where $\rho_D$ is the Poincaré metric and $\rho$ is any pseudometric assigned to $\mathcal{D}$ by a Schwarz-Pick system. Therefore it is natural to define $\rho_C$: $\mathcal{D} \times \mathcal{D} \to R$ by

$$\rho_C(x,y) = \sup \{\rho_D(f(x),f(y)): f \in \mathrm{Hol}(\mathcal{D},D)\}.$$

It is not difficult to see that $\rho_C$ is a pseudometric and that these pseudometrics constitute a Schwarz-Pick system. Of all Schwarz-Pick systems, this system (called Carathéodory's system) assigns the smallest pseudometric to a given domain.

If x and y are in $\mathcal{D}$ and there is a function $f \in \mathrm{Hol}(D, \mathcal{D})$ such that both x and y belong to f(D), then x = f(z) and y = f(w) for some $z,w \in D$, and

$$\rho(x,y) = \rho(f(z),f(w)) \leq \rho_D(z,w) \tag{11.4}$$

for any pseudometric $\rho$ assigned to $\mathcal{D}$ by a Schwarz-Pick system. Therefore we define

$$\rho_K^*(x,y) = \inf\{\rho_D(z,w): x = f(z) \text{ and } y = f(w) \text{ for some holomorphic } f: D \to \mathcal{D}\}.$$

Given x in $\mathcal{D}$, $\rho_K^*(x\ y)$ is defined for all y sufficiently close to x. Clearly $\rho(x,y) \leq \rho_K^*(x,y)$ whenever the right-hand side is defined. We now set

$$\rho_K(x,y) = \inf \{ \sum_{j=1}^{n} \rho_K^*(x_{j-1}, x_j)\}$$

where the infimum is taken over all positive integers n and all chains $x = x_0,x_1,x_2,\ldots,x_n = y$ for which $\rho_K^*(x_{j-1}, x_j)$ is defined for all $1 \leq j \leq n$. (Such a chain always exists because $\mathcal{D}$ is connected.) Once again it is not difficult to see that $\rho_K$ is a pseudometric and that these pseudometrics constitute a Schwarz-Pick system. Of all Schwarz-Pick systems, this system (called Kobayashi's system) assigns the largest pseudometric to a given domain.

Thus we see that

$$\rho_C(x,y) \leqslant \rho(x,y) \leqslant \rho_K(x,y) \tag{11.5}$$

for any $\rho$ assigned to $\mathcal{D}$ by a Schwarz-Pick system. In particular,

$$\rho_C(x,y) \leqslant \rho_{CRF}(x,y) \leqslant \rho_K(x,y) \ .$$

Strict inequalities hold for some domains [48].

Note that any pseudometric assigned to a Banach space X by a Schwarz-Pick system vanishes identically. It follows that if at least one Schwarz-Pick system assigns a metric to a domain $\mathcal{D}$ , then $\mathrm{Hol}(X, \mathcal{D})$ consists only of constant maps.

By (11.5), such a metric exists if and only if the Kobayashi pseudometric $\rho_K$ assigned to $\mathcal{D}$ is a metric. We remark in passing that a domain for which $\rho_K$ is a metric equivalent to the norm metric is said to be hyperbolic. For example, any bounded domain is hyperbolic. See [36] and [48] for more details.

## 12. AUTOMORPHISMS AND CARTAN'S UNIQUENESS THEOREM

Let $\mathcal{D}_1 \subset X_1$ and $\mathcal{D}_2 \subset X_2$ be two domains contained in the Banach spaces $X_1$ and $X_2$. A mapping $f \in \mathrm{Hol}(\mathcal{D}_1, \mathcal{D}_2)$ is said to be biholomorphic if $f(\mathcal{D}_1) = \mathcal{D}_2$, f is one-to-one, and $f^{-1} \in \mathrm{Hol}(\mathcal{D}_2, \mathcal{D}_1)$. If such a mapping f exists, then $\mathcal{D}_1$ and $\mathcal{D}_2$ are said to be (holomorphically) equivalent. In this case,

$$\rho_2(f(x),\ f(y)) = \rho_1(x,y)$$

for each pair $\rho_1$ and $\rho_2$ of pseudometrics assigned to $\mathcal{D}_1$ and $\mathcal{D}_2$ respectively by the same Schwarz-Pick system, and for all x and y in $\mathcal{D}_1$.

Biholomorphic mappings of a domain $\mathcal{D}$ onto itself are called automorphisms. The group of all such mappings is denoted by $\mathrm{Aut}(\mathcal{D})$. A domain $\mathcal{D}$ is said to be homogeneous if for any two points x and y in $\mathcal{D}$ there exists an automorphism h in $\mathrm{Aut}(\mathcal{D})$ with $h(x) = y$. We already know that the open unit disc is a homogeneous domain. So are the open unit balls of some (but not all) Banach spaces. If such a ball $B(0,1)$ is homogeneous, then all the Schwarz-Pick systems coincide on it. Indeed if $x,y \in B(0,1)$ and $h \in \mathrm{Aut}(B(0,1))$ satisfies $h(x) = 0$, then for any pseudometric $\rho$,

$$\rho(x,y) = \rho(h(x),h(y)) = \rho(0, h(y)) = \operatorname{argtanh} |h(y)|.$$

It follows that any two points of $B(0,1)$ can be joined by a metric segment (the image under $h^{-1}$ of the Euclidean segment joining 0 and $h(y)$). Such a segment, however, need not be unique.

In order to obtain more information on $\mathrm{Aut}(B(0,1))$ we now establish Cartan's uniqueness theorem for bounded domains. Let I denote the identity mapping.

Theorem 12.1. Let $\mathcal{D}$ be a bounded domain in a Banach space X, and $f\colon \mathcal{D} \to \mathcal{D}$ a holomorphic function. If $f(x_0) = x_0$ and $Df(x_0) = I$ for some $x_0$ in $\mathcal{D}$, then $f = I$.

Proof. Without loss of generality we may assume that $x_0$ is the origin. Let $B(0,r) \subset \mathcal{D} \subset B(0,M)$. By Taylor's Theorem (Theorem 9.1),

$$f(x) = x + \sum_{k=2}^{\infty} P_k(x)$$

for all x in $B(0,r)$, where $P_k$ is the "homogeneous polynomial"

$$P_k(x) = \frac{1}{k!} \hat{D}^k f(0)x.$$

If the theorem were false, then the identity theorem (Theorem 9.4) would show that there exists a smallest integer $q \geq 2$ such that $P_q$ does not vanish identically. The Taylor expansion of $f^n$ (the n-th iterate of f) would then begin with $x + n\,P_q(x)$, and Cauchy's estimates would show that

$$|nP_q(x)| \leq \frac{M|x|^q}{r^q} < M.$$

Hence $P_q$ does vanish identically. This contradiction completes the proof.

We now use Cartan's uniqueness theorem to prove a result concerning automorphisms of the unit ball in a Banach space X.

Theorem 12.2. If $h \in \mathrm{Aut}(B(0,1))$ and $h(0) = 0$, then h is the restriction to $B(0,1)$ of a linear isometry of X onto itself.

Proof. We first observe that in this case $Dh(0)$ is an isometry of X onto itself with $(Dh(0))^{-1} = Dh^{-1}(0)$. Using the notation of the proof of Theorem 12.1, we write $h(x) = P_1(x) + P_2(x) + \dots$ . Let $\phi \in R$. The function $g\colon B(0,1) \to B(0,1)$ defined by $g(x) = h^{-1}(e^{-i\phi}h(e^{i\phi}x))$ is an automorphism. Since $g(0) = 0$ and $Dg(0) = I$, Theorem 12.1 shows that $g = I$. Hence

$$e^{i\phi}h(x) = h(e^{i\phi}x)$$

and

$$e^{i\phi}(P_1(x) + P_2(x) + \dots) = e^{i\phi}P_1(x) + e^{2i\phi}P_2(x) + \dots$$

for all $x \in B(0,1)$ and $\phi \in R$. It follows that $P_k(x) = 0$ for all $k \geq 2$, so that $h(x) = P_1(x) = Dh(0)x$, as claimed.

Thus we see that an automorphism of the unit ball that fixes the origin is not only an isometry with respect to $\rho$, but also with respect to the norm.

## 13. A FIXED POINT THEOREM

Let $\mathcal{D}$ be a domain in a Banach space X. A subset K of $\mathcal{D}$ is said to lie strictly inside $\mathcal{D}$ if $\mathrm{dist}(K,\partial\mathcal{D}) > 0$ (equivalently, if there exists a positive r such that $B(x,r) \subset \mathcal{D}$ for all x in K). A mapping $f\colon \mathcal{D} \to \mathcal{D}$ is said to map $\mathcal{D}$ strictly inside $\mathcal{D}$ if $f(\mathcal{D})$ lies strictly inside $\mathcal{D}$.

If K lies strictly inside $\mathcal{D}$ then so does its norm closure $\overline{K}$. If $\mathcal{D}$ is bounded, then any closed subset C of $\mathcal{D}$ lying strictly inside $\mathcal{D}$ is complete with respect to the restriction to C of any metric assigned to $\mathcal{D}$ by a Schwarz-Pick system. (See Sections 10 and 11.)

Theorem 13.1. Let $\mathcal{D}$ be a bounded domain in a Banach space X, and let $\rho$ be the CRF metric on $\mathcal{D}$. If a holomorphic $f\colon \mathcal{D} \to \mathcal{D}$ maps $\mathcal{D}$ strictly inside itself, then there exists $0 \leq k < 1$ such that $\rho(f(x),f(y)) \leq k\rho(x,y)$ for all x and y in $\mathcal{D}$.

Proof. Choose $r > 0$ such that $B(f(x),r) \subset \mathcal{D}$ for all x in $\mathcal{D}$, and let d be the diameter of $\mathcal{D}$. Set $t = r/d$, fix $x \in \mathcal{D}$, and define a mapping g in $\mathrm{Hol}(\mathcal{D},\mathcal{D})$ by $g(u) = f(u) + t(f(u) - f(x))$.

Lemma 10.1 shows that

$$\alpha(g(u),\ Dg(u)v) \leq \alpha(u,v)$$

for all u in $\mathcal{D}$ and v in X. Set $u = x$. Since $Dg = (1 + t)Df$, it follows that $\alpha(f(x),\ Df(x)v) \leq \alpha(x,v)/(1 + t)$. Since this inequality holds for all $x \in \mathcal{D}$, we conclude that $\rho(f(x),\ f(y)) \leq \rho(x,y)/(1 + t)$ for all x and y in $\mathcal{D}$. The proof is complete.

Let C be the norm closure of $f(\mathcal{D})$. Applying Banach's fixed point theorem to the restriction of f to C, we obtain the following fixed point theorem which is due to Earle and Hamilton [29].

Theorem 13.2. Let $\mathcal{D}$ be a bounded domain in a Banach space. If a holomorphic $f\colon \mathcal{D} \to \mathcal{D}$ maps $\mathcal{D}$ strictly into itself, then it has a unique fixed point. Moreover, for any x in $\mathcal{D}$ the sequence of iterates $\{f^n(x)\}$ converges to this fixed point.

## 14. THE AUTOMORPHISMS OF THE HILBERT BALL

We now begin a detailed study of the open unit ball $B = B(0,1)$ of a (complex) Hilbert space H of dimension $\geq 2$.

In the one-dimensional case, the Möbius transformation

$$m_a(z) = \frac{z + a}{1 + z\bar{a}}$$

belongs to Aut(D) for all a in D. It seems that the analog of this mapping for the case $\dim(H) \geq 2$ should be the function

$$m_a(z) = \frac{z + a}{1 + (z,a)},$$

where $(z,a)$ denotes the inner product of z and a. However, if z is orthogonal to a, then $m_a(z) = z + a$, so that $m_a$ does not map B back into itself (unless $a = 0$).

In order to find the correct analog of the one-dimensional Möbius transformation, let $P_a$ be the orthogonal projection of H onto the one-dimensional subspace spanned by a,

$$P_a(z) = \frac{(z,a)a}{|a|^2},$$

and let $Q_a = I - P_a$.

We have

$$P_a m_a(z) = \frac{P_a z + a}{1 + (z,a)}, \quad \text{and}$$

$$Q_a m_a(z) = \frac{Q_a z}{1 + (z,a)}.$$

Since $|P_a m_a(z)|^2 + (1 - |a|^2)|Q_a m_a(z)|^2 < 1$ for all z in B, we can define, for each $a \in B$, a holomorphic mapping $M_a : B \to B$ by

$$M_a(z) = (\sqrt{1 - |a|^2}\, Q_a + P_a)m_a(z) .$$

Since $M_{-a} \circ M_a = I$, we see that $M_a$ belongs to Aut(B). We shall call $M_a$ a Möbius transformation. Note that $M_a(0) = a$ and that $M_{-a}(a) = 0$. Therefore $M_b \circ M_{-a}(a) = b$ and we conclude that the group Aut(B) acts transitively on B.

The following result describes the whole group of automorphisms of B.

Theorem 14.1. If $h \in \mathrm{Aut}(B)$, then $h = U \circ M_a$ for some unitary operator U and a in B.

Proof. Let $a = -h^{-1}(0)$. Then $g = h \circ M_{-a}$ belongs to Aut(B) and $g(0) = 0$. Therefore g is the restriction to B of a unitary operator $U: H \to H$ by Theorem 12.2. The result follows.

This result shows that any automorphism h in Aut(B) has, in fact, a (norm) continuous extension to the closed unit ball $\overline{B}$. This is because it is the restriction to B of a holomorphic function defined on the larger ball B(0,R) where $R = 1/|h^{-1}(0)|$.

It is of interest to mention that if $a \neq 0$, then both the (metric) segment $\{ta/|a| : -1 < t < 1\}$ and the one-dimensional disc $\{\xi \frac{a}{|a|} : |\xi| < 1\}$ are invariant under $M_a$.

Finally, we note that Aut(B) also contains all mappings of the form $S_a(z) = M_a(-M_{-a}(z))$. These automorphisms satisfy $S_a^2 = I$ and have exactly one fixed point, namely a.

## 15. THE HYPERBOLIC METRIC ON THE HILBERT BALL

All Schwarz-Pick systems assign the same metric to the Hilbert unit ball B. We shall call this unique metric the hyperbolic metric and denote it by $\rho$. We know that

$$\rho(x,y) = \operatorname{argtanh} |M_{-x}(y)|.$$

Therefore a computation shows that

$$\rho(x,y) = \operatorname{argtanh} (1-\sigma(x,y))^{1/2},$$

where

$$\sigma(x,y) = \frac{(1 - |x|^2)(1 - |y|^2)}{|1 - (x,y)|^2}.$$

Note that $\rho(x,y) \leq \rho(u,v)$ if and only if $\sigma(x,y) \geq \sigma(u,v)$, and that $x = y$ if and only if $\sigma(x,y) = 1$.

In contrast with the one-dimensional case, balls in $(B,\rho)$ are no longer balls in the sense of the norm. In order to describe the ball $B(a,r) = \{x \in B: \rho(a,x) < r\}$, we introduce the notations

$$d = \tanh r, \quad k = \frac{1 - |a|^2}{1 - d^2}, \quad \text{and}$$

$$\phi_a(x) = |1 - (x,a)|^2/(1 - |x|^2).$$

It is clear that the following inequalities are equivalent:

$$\rho(a,x) < r; \tag{15.1}$$

$$|M_{-a}(x)| < d; \tag{15.2}$$

$$\sigma(a,x) > 1 - d^2; \tag{15.3}$$

$$\phi_a(x) < k. \tag{15.4}$$

Using the orthogonal projections $P_a$ and $Q_a$, we can rewrite these inequalities in the form

$$\frac{|P_a x - c|^2}{d^2 m^2} + \frac{|Q_a x|^2}{d^2 m} < 1, \tag{15.5}$$

where $c = \dfrac{(1 - d^2)a}{1 - d^2|a|^2}$ and

$$m = \frac{1 - |a|^2}{1 - d^2|a|^2}.$$

Thus B(a,r) is an "ellipsoid" with center at c. The intersection of this ellipsoid with the plane spanned by a (or c) is a disc of radius dm and its intersection with a plane orthogonal to a at c is a disc with the "much larger" radius $d\sqrt{m}$.

Still another representation of B(a,r) may be obtained by using coordinates. Set $e = a/|a|$ and represent each x in B as $x = \xi e + x'$ where $\xi = (x,e)$ and $(x',e) = 0$. Then x may be represented as $(\xi,x')$ and the ball B(a,r) may be described by the inequality

$$\frac{|\xi - |c||^2}{d^2 m^2} + \frac{|x'|^2}{d^2 m} < 1. \tag{15.6}$$

The most useful representation of balls is perhaps (15.4). For any $k > 1 - |a|^2$, the inequality (15.4) represents the ball centered at a with radius

$$r = \operatorname{argtanh}\left(1 - \frac{1 - |a|^2}{k}\right)^{1/2}.$$

Thus $r \to 0$ as $k \to 1 - |a|^2$ and $r \to \infty$ as $k \to \infty$.

Observe, however, that $\phi_a(x)$ is well-defined for all x in B even if $|a| = 1$. In this case (15.4) makes sense for all $k > 0$. Let then e belong to the boundary of B and consider the set of all $x \in B$ such that

$$\phi_e(x) < k. \tag{15.7}$$

This inequality is equivalent to

$$\frac{|P_e x - (1-\alpha)e|^2}{\alpha^2} + \frac{|Q_e x|^2}{\alpha} < 1, \tag{15.8}$$

where $k = \alpha/(1-\alpha)$ .

Thus we see that $\{x \in B: \phi_e(x) < k\}$ is again an ellipsoid. Using coordinates, we can rewrite (15.8) in the form

$$\frac{|\xi - (1-\alpha)|^2}{\alpha^2} + \frac{|x'|^2}{\alpha} < 1, \tag{15.9}$$

where $x = (\xi, x')$.

This ellipsoid is $\rho$-unbounded and therefore is no longer a $\rho$-ball. In fact, it may be considered a half-space (of the first kind). To see this, let $0 \leq t < 1$ and consider the $\rho$-balls $B(se, \rho(se,te))$ with $t < s < 1$ (cf. Section 3). When $s \to 1$, we obtain the ellipsoid $\{x \in B: \phi_e(x) < k\}$ with $k = (1-t)/(1+t)$. Note that $\alpha = (1-t)/2$ , just as in the one-dimensional case. (Such an ellipsoid is sometimes called a horosphere.)

In view of these facts, we introduce the notation

$$E(a,k) = \{x \in B: \phi_a(x) < k\}$$

for any $a \in \overline{B}$ and $k > 1 - |a|^2$. This set is either a $\rho$-ball (if $|a| < 1$) or a half-space of the first kind (if $|a| = 1$).

Finally we observe that the $\rho$-closure of $B(a,r) = E(a,k)$ is equal to its norm closure $\overline{E}(a,k)$, but that the $\rho$-closure of the half-space $E(a,k)$ ($|a| = 1$) equals $\overline{E}(a,k) \setminus \{a\}$. This is because in this case the norm closure of $E(a,k)$ intersects the boundary at a.

## 16. GEODESICS AND AFFINE SETS

We already know that for any point e of norm 1, the set $\{te:-1<t<1\}$ is a metric line passing through the origin. As a matter of fact, each geodesic passing through the origin is of this form. This is because the equality

$$\rho(x,0) + \rho(0,y) = \rho(x,y)$$

implies that $y = sx$ with $s \leq 0$. All the other geodesics in B are the images of these lines under the automorphisms of B. In order to describe them, we consider affine subsets of B.

We shall say that a subset A of B is affine if $A = B \cap L$ for some closed affine (complex) submanifold of H. In particular, if L is one-dimensional, then we shall call $B \cap L$ a complex line in B.

Let $e \in \partial B$ and let $D_e = \{\xi e : \xi \in D\}$ be a complex line passing through the origin. As before, represent any point $x \in B$ as $x = (\xi, x')$. Consider the mapping $f: B \to D$ defined by $f(x) = \xi$. Since $f \in \text{Hol}\,(B,D)$,

$$\rho_D(\xi,\eta) \leq \rho_B(\xi e . \eta e)$$

for all $\xi$ and $\eta$ in D. On the other hand, the embedding $g: D \to B$ defined by $g(\xi) = \xi e$ is also holomorphic and we have

$$\rho_B(\xi e,\eta e) \leq \rho_D(\xi,\eta) .$$

Thus the complex line $D_e$ is isometric to D. Also, the geodesics in $D_e$ are arcs of circles orthogonal to $\overline{D}_e \cap \partial B$.

Now consider the geodesic $\Gamma$ joining the points z and w in B. These points determine a complex line $A = L \cap B$ in B. Let $a \in B$ be the unique point of least norm in L. Let $x \neq a$ be any other point in L. Denoting $(x-a)/|x-a|$ by e, we may represent any point $y \in A$ as

$$y = a + \xi\sqrt{1 - |a|^2}\, e$$

with $\xi$ in D.

The Möbius transformation $M_{-a}$ maps A onto $D_e$. Thus A is isometric to D and the geodesics joining two points in A (including $\Gamma$ of course) are arcs of circles contained in A and orthogonal to $L \cap \partial B$. (Geodesics joining a and $y \in A$ are Euclidean segments.)

The same reasoning shows that if L is an n-dimensional affine manifold, then the affine set $L \cap B$ is isometric to the n-dimensional Hilbert ball $B_n$. Actually, even more may be said. Note that the image of an affine set under any automorphism of B is again an affine set. Also, any affine set may be transformed by a Möbius transformation onto an affine set containing the origin. Finally note that for any two affine sets of the same dimension that contain the origin there is a unitary operator mapping one onto the other. Hence two affine sets are isometric if and only if they are of the same dimension (finite or infinite). It is also clear that any geodesic may be transformed onto any other by an appropriate automorphism.

## 17. ρ-CONVEXITY

In view of the uniqueness of geodesics we can define metric centers and ρ-convex combinations exactly as we did in the one-dimensional case (Section 4). For any x and y in B and $0 \leq t \leq 1$ we let $(1-t)x \oplus ty$ stand for the unique point z satisfying

$$\rho(x,z) = t\rho(x,y) \text{ and } \rho(z,y) = (1-t)\,\rho(x,y).$$

The metric segment joining x and y will be denoted by $[x,y]$. A subset K of B is said to be ρ-convex if $[x,y] \subset K$ for all x and y in K. As in the one-dimensional case. the family of ρ-convex subsets of B satisfies properties (a), (b), (c) and (d) of Section 4.

(The analog of Property (e) will be discussed in Section 20.) It is obvious that all affine sets are $\rho$-convex.

Lemma 17.1. If a,b,x and y are in B and $0 \leq t \leq 1$, then $\rho((1-t)a \oplus tx, (1-t)a \oplus ty) \leq t\rho(x,y)$, and

$$\rho((1-t)a \oplus tx, (1-t)b \oplus ty) \leq (1-t)\rho(a,b) + t\rho(x,y).$$

Strict inequalities occur if $0 < t < 1$ and the relevant points do not lie on the same geodesic.

Proof. It suffices to prove the first inequality for $t = 1/2$ and $a = 0$ and then repeat the reasoning used in the one-dimensional case (Section 6). To this end, let

$$u = \frac{1}{2} x \oplus \frac{1}{2} 0$$

and

$$v = \frac{1}{2} y \oplus \frac{1}{2} 0.$$

Then $x = 2u/(1 + |u|^2)$ and $y = 2v/(1 + |v|^2)$. The claimed inequality

$$\rho(u,v) \leq \frac{1}{2} \rho(x,y)$$

is equivalent to

$$\sigma(x,y) \leq \frac{\sigma^2(u,v)}{(2 - \sigma(u,v))^2}. \tag{17.1}$$

To establish (17.1), we compute to find that

$$\sigma(x,y) = \frac{(1 - |u|^2)^2 (1 - |v|^2)^2}{|(1 - |u|^2)(1 - |v|^2) + 2(|u|^2 + |v|^2 - 2(u,v))|^2}$$

$$\leqslant \frac{(1-|u|^2)^2(1-|v|^2)^2}{((1-|u|^2)(1-|v|^2)+2\,|u-v|^2)^2},$$

with strict inequality if $\operatorname{Im}(u,v) \neq 0$.

In other words,

$$\sigma(x,y) \leqslant \frac{\sigma^2(u,v)}{\left(\sigma(u,v)+\dfrac{2|u-v|^2}{|1-(u,v)|^2}\right)^2}.$$

Now it is enough to observe that the inequality

$$\sigma(u,v)+\frac{2|u-v|^2}{|1-(u,v)|^2} \geqslant 2-\sigma(u,v)$$

is equivalent to $|(u,v)| \leqslant |u|\,|v|$.

It follows that (17.1) is indeed true, with equality if and only if x,y and the origin lie on the same geodesic. This completes the proof.

## 18. UNIFORM CONVEXITY OF $(B,\rho)$

As in the one-dimensional case, we define the modulus of convexity $\delta:(0,\infty)\times(0,2]\to(0,1]$ of $(B,\rho)$ by

$$\delta(r,\varepsilon) = \inf\{1-\frac{1}{r}\,\rho(a,\frac{1}{2}x\oplus\frac{1}{2}y)\},$$

where the infimum is taken over all points a,x and y in B satisfying $\rho(a,x) \leqslant r$, $\rho(a,y) \leqslant r$, and $\rho(x,y) \geqslant \varepsilon r$.

In order to compute $\delta$, let $u = \frac{1}{2}x\oplus\frac{1}{2}y$ and apply $M_{-u}$ to a, x and y. Set $b = M_{-u}(a)$ and $w = M_{-u}(x) = -M_{-u}(y)$. Then we have

$$r \geqslant \rho(a,x) = \rho(b,w) = \operatorname{argtanh}\,(1-\sigma(b,w))^{1/2},$$

and

$$r \geq \rho(a,y) = \rho(b,-w) = \operatorname{argtanh}\,(1 - \sigma(b,-w))^{1/2}.$$

Hence

$$\sigma(b,w) = \frac{(1 - |b|^2)(1 - |w|^2)}{|1 - (b,w)|^2} \geq 1 - \tanh^2 r,$$

and

$$\sigma(b,-w) = \frac{(1 - |b|^2)(1 - |w|^2)}{|1 + (b,w)|^2} \geq 1 - \tanh^2 r.$$

At least one of the denominators of the left-hand sides of these inequalities must exceed 1. We also have

$$|w| = \tanh \rho(0,w) = \tanh\,(\tfrac{1}{2}\,\rho(x,y)) \geq \tanh\,(\tfrac{1}{2}\,\varepsilon r).$$

Therefore

$$|b|^2 \leq \frac{\tanh^2 r - \tanh^2(\tfrac{1}{2}\,\varepsilon r)}{1 - \tanh^2(\tfrac{1}{2}\,\varepsilon r)}$$

$$= \frac{\sinh(r(1 + \varepsilon/2))\,\sinh(r(1 - \varepsilon/2))}{\cosh^2 r},$$

so that

$$\rho(a, \tfrac{1}{2}\,x \oplus \tfrac{1}{2}\,y) = \rho(0,b)$$

$$\leq \operatorname{argtanh} \frac{[\sinh(r(1 + \varepsilon/2))\,\sinh(r(1 - \varepsilon/2))]^{1/2}}{\cosh r}$$

Since these inequalities are the best possible (equalities occur if a,x and y are chosen so that $(b,w) = 0$), we conclude that

$$\delta(r,\varepsilon) = 1 - \frac{1}{r}\operatorname{argtanh}\frac{[\sinh(r(1+\varepsilon/2)]\sinh(r(1-\varepsilon/2))]^{1/2}}{\cosh r}.$$

Although the modulus of convexity of $(B,\rho)$ is smaller than the modulus of convexity of $(D,\rho)$, it does satisfy properties (a), (b), (c) and (d) of Section 5. In addition, it is obvious that the following implication is once again valid:

$$\left.\begin{array}{l}\rho(a,x) \leq r\\ \rho(a,y) \leq r\\ \rho(x,y) \geq \varepsilon r\end{array}\right\} \Rightarrow \rho(a, \tfrac{1}{2}x \oplus \tfrac{1}{2}y) \leq (1 - \delta(r,\varepsilon))r.$$

Thus we see again that the rotundity of small balls in $(B,\rho)$ is more or less the same as the rotundity of balls in Hilbert space, while large balls are almost square.

If $\dim H \geq 2$, then the balls in $(B,\rho)$ are less uniformly rotund than balls in the unit disc.

A basic consequence of uniform convexity is the following intersection theorem. We omit the proof because it is essentially identical to the proof of the corresponding result in uniformly convex Banach spaces (Theorem 2.1 of Chapter 1). This theorem is also valid for arbitrary (not necessarily countable) decreasing family of $\rho$-convex sets.

Theorem 18.1. Let $\{K_n : n = 1,2,\ldots\}$ be a decreasing sequence of nonempty $\rho$-bounded closed $\rho$-convex subsets of B. Then the intersection $\bigcap\{K_n : n = 1,2,\ldots\}$ is a nonempty closed $\rho$-convex subset of B.

Let K be a $\rho$-convex subset of $(B,\rho)$. We shall say that a function $f\colon K \to (-\infty,\infty)$ is $\rho$-convex if

$$f((1-t)x \oplus ty) \leq (1-t)f(x) + tf(y) \tag{18.1}$$

for all x and y in K and all $0 \leq t \leq 1$. It will be called strictly $\rho$-convex if strict inequality holds in (18.1) for $x \neq y$ and $0 < t < 1$. This implies that

$$f(\tfrac{1}{2}x \oplus \tfrac{1}{2}y) < \max\{f(x),f(y)\} \text{ for all } x \neq y. \tag{18.2}$$

As in the Banach space case, the following useful result is a direct consequence of Theorem 18.1.

Proposition 18.2. Let K be a $\rho$-closed, $\rho$-convex subset of $(B,\rho)$, and let $f\colon K \to [0,\infty)$ be a $\rho$-convex function. If f is lower semicontinuous and $f(x) \to \infty$ as $\rho(0,x) \to \infty$, then f attains its minimum on K. If, in addition, f satisfies (18.2), then it attains its minimum at exactly one point.

## 19. NEAREST POINT PROJECTIONS

We shall say that a subset D of $(B,\rho)$ is a $\rho$-Chebyshev set if to each point x in B there corresponds a unique point z in D such that $\rho(x,z) = \inf\{\rho(x,y)\colon y \in D\}$. In this case we define the nearest point projection $R\colon B \to D$ by assigning z to x. The following result is an application of Proposition 18.2.

Theorem 19.1. Every $\rho$-closed, $\rho$-convex subset of $(B,\rho)$ is a $\rho$-Chebyshev set.

Proof. Let K be a $\rho$-closed, $\rho$-convex subset of $(B,\rho)$, and let x belong to B. The function $f\colon K \to [0,\infty)$ defined by $f(y) = \rho(x,y)$ for y in K is $\rho$-convex by Lemma 17.1. Since $(B,\rho)$ is uniformly convex, it also satisfies (18.2). Therefore this result is indeed a consequence of Proposition 18.2.

Consider now the nearest point projection $R = R_A$ onto an affine set A in B. Assume first that $0 \in A$. Then $A = B \cap E$ where E is a closed subspace of H. Let $P = P_E : H \to E$ be the orthogonal projection onto E. For any $x \in B$ and $y \in A$ we have

$$\sigma(x,y) = \frac{(1 - |x|^2)(1 - |y|^2)}{|1 - (x,y)|^2}$$

$$= \frac{(1 - |x|^2)(1 - |y|^2)}{|1 - (Px,y)|^2} = \frac{1 - |x|^2}{1 - |Px|^2}\,\sigma(Px,y).$$

Hence

$$\sup\{\sigma(x,y) : y \in A\} = \frac{1 - |x|^2}{1 - |Px|^2}$$

is attained for $y = Px$ and $R_A(x) = P_E(x)$.

If the origin does not belong to A, then A can be represented as $(a + E) \cap B$, where a is the unique point of least norm in A and E is again a closed subspace of H. It follows that in this case $R_A = M_a \circ P_E \circ M_{-a}$, or more explicitly,

$$R_A(x) = a + \frac{1 - |a|^2}{1 - (Q_E x, a)} P_E(x),$$

where $x \in B$ and $Q_E = I - P_E$.

Thus in both cases $R_A$ is holomorphic, hence ρ-nonexpansive.

Next we consider the nearest point projection onto a metric line. Assume first that the metric line γ passes through the origin. Then it is of the form $(-1,1)e$, where $|e| = 1$. Let A be the complex line determined by e. Let $R = R_A$ be the nearest point projection from B onto A. Since

$$\sigma(x,te) = \frac{1 - |x|^2}{1 - |Rx|^2}\, \sigma(Rx,te),$$

we see that $R_\gamma(x) = R_\gamma(R_A(x))$. We already know by Theorem 6.10 that the restriction of $R_\gamma$ to A is ρ-nonexpansive. Therefore this equality shows that $R_\gamma$ is ρ-nonexpansive on all of B. The homogeneity of B now shows that the nearest point projection onto any metric line is ρ-nonexpansive. Consequently, we can use the method of proof of Theorem 6.10 to obtain the following general result.

Theorem 19.2. The nearest point projection onto a ρ-closed ρ-convex subset of (B,ρ) is ρ-nonexpansive.

The nearest point projection R onto the closed ρ-ball $\overline{B}(a,r)$ can also be found explicitly. It is clear $Rx = x$ if $x \in \overline{B}(a,r)$. If $x \notin \overline{B}(a,r)$ let $t = r/\rho(a,x)$ and $z = (1-t)a \oplus tx$. We have

$$\rho(a,x) \leq \rho(a,Rx) + \rho(Rx,x) = r + \rho(Rx,x).$$

Hence $\rho(Rx,x) \geq \rho(a,x) - r = \rho(x,z)$ and $Rx = z$. In other words, R is "radial" in this case.

We conclude this section by noting several other properties of the nearest point projection $R = R_K$ onto an arbitrary ρ-closed ρ-convex subset K of B.

For $0 \leq t \leq 1$, let $z = (1-t)x \oplus tRx$ and $y = Rz$. Then

$$\rho(x,y) \leq \rho(x,z) + \rho(z,y)$$

$$\leq \rho(x,z) + \rho(z,Rx) = \rho(x,Rx).$$

Hence $y = Rx$. In other words,

$$R((1-t)x \oplus tRx) = Rx$$

for all $0 \leq t \leq 1$. Now let $y$ be any point in $K$. Since

$$\rho(y,Rx) = \rho(Ry,R((1-t)x \oplus tRx))$$

$$\leq \rho(y,(1-t)x \oplus tRx) \leq (1-t)\rho(y,x) + t\rho(y,Rx),$$

we see that $\rho(y,Rx) \leq \rho(y,x)$. It follows that

$$\rho(y,Rx) \leq \rho(y,(1-t)x \oplus tRx) \text{ for all } 0 \leq t \leq 1.$$

In other words, the convex function $\phi:[0,1] \to [0,\infty)$ defined by $\phi(t) = \rho(y,(1-t)x \oplus tRx)$ attains its minimum at 1. Hence $\phi$ decreases on $[0,1]$ (strictly if $x \notin K$). We also see that $R_{[x,Rx]}(y) = Rx$. Since

$$\rho(x,Rx) \leq \rho(x,(1-t)y \oplus tRx) \text{ for all } 0 \leq t \leq 1,$$

the convex function $\psi$: $[0,1] \to [0,\infty)$ defined by

$$\psi(t) = \rho(x,(1-t)y \oplus tRx)$$

also decreases on $[0,1]$ (strictly if $y \neq Rx$).

Finally, observe that for any two points $x$ and $z$ in $B$,

$$\rho(Rx,Rz) = \rho(R((1-t)x \oplus tRx), R((1-t)z \oplus tRz))$$

$$\leq \rho((1-t)x \oplus tRx, (1-t)z \oplus tRz).$$

It follows that the convex function $\rho((1-t)x \oplus tRx, (1-t)z \oplus tRz)$ is non-increasing for $0 \leq t \leq 1$. This fact is analogous to the firm nonexpansiveness of the nearest point projections in Hilbert space. We shall take a closer look at the notion of firm nonexpansiveness in $(B,\rho)$ later.

## 20. EQUIDISTANT SETS AND HALF-SPACES

If x and y are two distinct points in B, we define the equidistant set $Eq(x,y)$ by $Eq(x,y) = \{z \in B: \rho(z,y) = \rho(z,x)\}$. As we did in the one-dimensional case, we also define two closed half-spaces (of the second kind) by

$$Eq^{+}(x,y) = \{z \in B: \rho(z,y) \leq \rho(z,x)\}$$

and $Eq^{-}(x,y) = Eq^{+}(y,x)$.

Let us first consider the special case $y = -x$. In this case, $Eq(x,y) = \{z \in B: Re(z,x) = 0\}$ and $Eq^{+}(x,y) = \{z \in B: Re(z,x) \leq 0\}$.

Let $\{e_\alpha: \alpha \in A\}$ be an orthonormal basis in H and let $\{\zeta_n\}$ and $\{\xi_n\}$ be the coordinates of z and x respectively. Writing $\zeta_n = \alpha_n + i\beta_n$ and $\xi_n = \gamma_n + i\delta_n$ with $\alpha_n, \beta_n, \gamma_n$ and $\delta_n$ real, we see that $Re(z,x) = \Sigma\zeta_n\bar{\xi}_n = \Sigma(\alpha_n\gamma_n + \beta_n\delta_n)$. Therefore we can identify $Eq(x,-x)$ with a real subspace of H of real codimension 1 which is orthogonal (in the real sense) to x.

Any other equidistant set $Eq(x,y)$ consists of all $z \in B$ that satisfy $Re(M_{-u}(z), M_{-u}(x)) = 0$ with $u = \frac{1}{2}x \oplus \frac{1}{2}y$. Therefore it may be described as the image under $M_u$ of $Eq(w,-w)$ where $w = M_{-u}(x)$.

In the one-dimensional case, $Eq(x,y)$ is the geodesic passing through $\frac{1}{2}x \oplus \frac{1}{2}y$ and orthogonal to the geodesic joining x and y. Therefore it (and the half-spaces determined by it) are ρ-convex. We show now that this is no longer true if $\dim H \geq 2$. (Cf. property (e) in Section 4.)

<u>Example 20.1.</u> Let $H = C^2$ and consider $Eq((i/2,0),(-i/2,0)) = \{(t,w) \in B: -1 < t < 1\}$. The complex line passing through $u = (i/2\sqrt{2}, i/2\sqrt{2})$ in the direction of the unit vector $e = (i/\sqrt{2}, -i/\sqrt{2})$ is the set of all points of the form $u + \xi e$ with $|\xi| < \sqrt{3}/2$ and is ρ-convex. The intersection of this complex line with $Eq((i/2,0), (-i/2,0))$ consists of all points of

the form $u + \xi e$ with $\mathrm{Re}\,\xi = -1/2$. Thus it is a Euclidean segment which does not include u. Hence it is not ρ-convex and neither is $\mathrm{Eq}((i/2,0), (-i/2,0))$.

As in the one-dimensional case, we now consider the set $A(x,y)$ consisting of all $z \in B$ for which the function

$$\phi(t) = \rho(z,(1-t)x \oplus ty)$$

is non-increasing on $[0,1]$. Let $2y \ominus x$ denote the unique point u on the metric line passing through x and y which satisfies

$$y = \frac{1}{2} x \oplus \frac{1}{2} u.$$

Since $\phi$ is convex, $z \in A(x,y)$ if and only if $\phi'(1) \leqslant 0$. This condition is equivalent to $\mathrm{Re}(M_{-y}(z), M_{-y}(x)) \leqslant 0$. Hence $A(x,y) = \overset{+}{\mathrm{Eq}}(x, 2y \ominus x)$. It is also clear that $z \in A(x,y)$ if and only if $\rho(z,y) = \min\{\phi(t): 0 \leqslant t \leqslant 1\}$.

Lemma 20.2. For any three points x,y and z in B the following are equivalent:

The function $\rho(y,(1-t)x \oplus tz)$ decreases on $[0,1]$; (20.1)

$$\rho(y,z) = \min\{\rho(y,(1-t)x \oplus tz): 0 \leqslant t \leqslant 1\}; \tag{20.2}$$

$$y \in \overset{+}{\mathrm{Eq}}(x, 2z \ominus x); \tag{20.3}$$

$$\mathrm{Re}(M_{-z}(y), M_{-z}(x)) \leqslant 0; \tag{20.4}$$

The function $\rho(x,(1-t)y \oplus tz)$ decreases on $[0,1]$; (20.5)

$$\rho(x,z) = \min\{\rho(x,(1-t)y \oplus tz): 0 \leqslant t \leqslant 1\}; \tag{20.6}$$

$$x \in \overset{+}{\mathrm{Eq}}(y, 2z \ominus y). \tag{20.7}$$

Proof. The discussion preceding the lemma shows that the two sets of conditions (20.1) - (20.4) and (20.4) - (20.7) are equivalent.

We now relate equidistant sets and half-spaces to the nearest point projection.

Theorem 20.3. Let $K$ be a $\rho$-convex $\rho$-closed subset of $B$, $R:B \to K$ the nearest point projection onto $K$, $z$ a point in $K$, and $x$ a point in $B$. Then the following are equivalent:

$$z = Rx; \tag{20.8}$$

$$\rho(y,z) = \min\{\rho(y, (1-t)x \oplus tz: 0 \leq t \leq 1\} \text{ for all } y \in K; \tag{20.9}$$

$$\operatorname{Re}\,(M_{-z}(y), M_{-z}(x)) \leq 0; \tag{20.10}$$

$$\rho(x,z) = \min\{\rho(x,(1-t)y \oplus tz): 0 \leq t \leq 1\} \text{ for all } y \in K; \tag{20.11}$$

$$K \subset \overset{+}{\mathrm{Eq}}(x, 2z \ominus x). \tag{20.12}$$

Proof. The last four conditions are all equivalent by Lemma 20.2. If (20.8) holds, then (20.11) follows immediately. Conversely, if (20.11) holds, then

$$\rho(x,z) \leq (1-t)\rho(x,y) + t\rho(x,z).$$

Hence $\rho(x,z) \leq \rho(x,y)$ for all $y \in K$ and $z = Rx$.

We conclude this section with another description of the half-spaces $\overset{+}{\mathrm{Eq}}(x,y)$. The inequality $\rho(z,y) \leq \rho(z,x)$ is equivalent to

$$\left| \frac{1 - (y,z)}{1 - (x,z)} \right| \leq \left( \frac{1 - |y|^2}{1 - |x|^2} \right)^{1/2} = c$$

which implies that

$$\frac{1 - (y,z)}{1 - (x,z)} = \xi , \tag{20.13}$$

where $\xi$ is a complex number such that $|\xi| \leq c$.

As in the one-dimensional case, (20.13) may be rewritten as

$$(x + \frac{d}{1-\xi} v,z) = 1, \tag{20.14}$$

where $d = |y-x|$ and $v = (y-x)/|y-x|$. In other words, z belongs to $\overset{+}{\mathrm{Eq}}(x,y)$ if and only if it satisfies (20.14) for some complex number $\xi$ with $|\xi| \leq c$.

Consider the one-dimensional plane L passing through x and y. Let a be the unique point of least norm in L. The affine set $L \cap B$ is a disc centered at a with radius $\sqrt{1 - |a|^2}$. The function $\xi \to \eta = x + \frac{d}{1-\xi} v$ maps the disc $|\xi| \leq c$ onto a disc, a half-plane, or the complement of a disc. In each case the image is invariant under inversion with respect to $L \cap B$. A point z belongs to $\overset{+}{\mathrm{Eq}}(x,y)$ if there is a point $p \notin B$ of the form $x + \frac{d}{1-\xi} v$ such that $(p,z) = 1$. Thus $z = p/|p|^2 + w$, where $(p,w) = 0$. If $a = 0$, this means that $\overset{+}{\mathrm{Eq}}(x,y)$ is a "cylinder" the intersection of which with $L \cap B$ is the one-dimensional half-space obtained in Section 7. In the general case $\overset{+}{\mathrm{Eq}}(x,y)$ is the image under $M_a$ of such a "cylinder".

## 21. ASYMPTOTIC CENTERS

Let $\{x_n\}$ be a ρ-bounded sequence in B, and let K be a ρ-closed ρ-convex subset of B. Consider the functional $f\colon B \to [0,\infty)$ defined by

$$f(x) = \limsup_{n\to\infty} \rho(x_n\, x).$$

A point $z$ in $K$ is said to be an asymptotic center of the sequence $\{x_n\}$ with respect to $K$ if $f(z) = \min\{f(x): x \in K\}$. The infimum of $f(x)$ over $K$ is called the asymptotic radius of $\{x_n\}$ with respect to $K$.

Proposition 21.1. Every $\rho$-bounded sequence in $(B,\rho)$ has a unique asymptotic center with respect to any $\rho$-closed $\rho$-convex subset of $B$.

Proof. Combine Proposition 18.2 with the method of proof of the analogous result in uniformly convex Banach space. (See Theorem 4.1 of Chapter 1.)

The asympotic center of $\{x_n\}$ with respect to $K$ will be denoted by $A(K,\{x_n\})$ and its asymptotic radius by $r(K,\{x_n\})$. If $K = B$ we shall write $A(\{x_n\})$ and $r(\{x_n\})$ respectively.

Lemma 21.2. If $\{x_n\} \subset K$, then $A(\{x_n\}) = A(K,\{x_n\})$.

Proof. Let $R = R_K: B \to K$ be the nearest point projection onto $K$. Since $r(Rx, \{x_n\}) \leq r(x, \{x_n\})$, $A(\{x_n\})$ must belong to $K$.

For any set $S \subset B$, let $\rho\text{-co}(S)$ denote the intersection of all $\rho$-convex sets containing $S$. Lemma 21.2 shows that

$$A(\{x_n\}) \in \bigcap_{n=1}^{\infty} \rho\text{-co}(\{x_k : k \geq n\}).$$

In this respect the asymptotic center is similar to the weak limit. Recall, however, that although the asymptotic center of a weakly convergent sequence in Hilbert space coincides with its weak limit, this assertion is not true in all uniformly convex Banach spaces. In order to relate $(B,\rho)$ to the weak topology of $B$ we first note the following fact.

Lemma 21.3. All the automorphisms of $(B,\rho)$ are weakly continuous.

Proposition 21.4. If a $\rho$-bounded sequence $\{x_n\}$ converges weakly to $x$, then $x = A(\{x_n\})$.

Proof. Since $A(\{h(x_n)\}) = h(A(\{x_n\}))$ for all $h \in \mathrm{Aut}(B)$, Lemma 21.3 shows that we may assume that $x = 0$. We also note that a sequence $\{x_n\}$ is $\rho$-bounded if and only if

$$\limsup_{n\to\infty} |x_n| < 1.$$

For any $y \neq 0$ we now have

$$\liminf \sigma(x_n, y) = \liminf (1 - |x_n|^2)(1 - |y|^2)/|1 - (x_n, y)|^2$$

$$= (1 - |y|^2) \liminf (1 - |x_n|^2) < \liminf \sigma(x_n, 0).$$

Therefore

$$r(0, \{x_n\}) < r(y, \{x_n\}) \text{ and } A(\{x_n\}) = 0.$$

Actually we can obtain even more accurate information.

Proposition 21.5. If a $\rho$-bounded sequence $\{x_n\}$ converges weakly to $x$, then $A(K, \{x_n\}) = R_K(x)$.

Proof. As in the proof of Proposition 21.4, we assume that $x = 0$. We then see that $\sup\{\liminf \sigma(x_n, y) : y \in K\}$ is attained when $y = R_K(0)$.

Corollary 21.6. A $\rho$-bounded, closed, and $\rho$-convex subset of $(B, \rho)$ is weakly compact.

Proof. Let $\{x_n\}$ be a weakly convergent sequence contained in such a subset K. Its weak limit $x = A(\{x_n\}) = A(K, \{x_n\})$ by Proposition 21.4 and Lemma 21.2. Therefore it must belong to K.

## 22. NONEXPANSIVE MAPPINGS IN $(B,\rho)$

As in the one-dimensional case, we denote by $N = N(B)$ the family of all $\rho$-nonexpansive mappings of B into itself, and by $N_k$ the subset of N consisting of all strict $\rho$-contractions with constant k. We already know that $\mathrm{Hol}(B) \subset N$ and that any $f \in \mathrm{Hol}(B)$ which maps B strictly inside B is a strict $\rho$-contraction. The family N is convex not only in the linear sense, but also in the metric sense. If $f \in N_k$, $g \in N_\ell$, and $0 \leq t \leq 1$, then $(1-t)f + tg$ belongs to $N_{\max(k,\ell)}$ and $(1-t)f \oplus tg$ belongs to $N_{(1-t)k+t\ell}$. It can also be shown that $\rho(kx,ky) \leq k\rho(x,y)$ for all $0 \leq k \leq 1$. N contains all the nearest point projections onto $\rho$-closed $\rho$-convex subsets of B. The anti-unitary mappings (that is, those linear operators U for which $(Ux,Uy) = \overline{(x,y)}$ for all x and y) also belong to N. It is obvious that the family N is closed with respect to composition. It is also closed with respect to pointwise convergence in the same sense as in the one-dimensional case.

Note, however, that some norm isometries of B are not $\rho$-nonexpansive. For example, if $H = C^2$, then $f:B \to B$ defined by $f(z,w) = (\bar{z},w)$ is such a mapping.

There is also a significant difference between $(B,\rho)$ and Hilbert space. Recall that if $P: H \to C$ is the (norm) nearest point projection onto a closed convex subset C of H, then the mapping $2P-I$ is nonexpansive. The following example shows that the analog of this fact is no longer true in $(B,\rho)$.

Example 22.1. Let $H = C^2$ and $K = \{(t,0): -1 < t < 1\}$. In this case, $R_K(x,y) = (t,0)$ where $t/(1+t^2) = \mathrm{Re}\,x/(1+|x|^2)$, and $S(x,y) = 2R_K(x,y) \ominus (x,y) = (\bar{x},-y)$. Since $S^2 = I$, S would be an isometry if it were $\rho$-nonexpansive. But a computation shows that

$$\sigma(S(x,y),\ S(u,v)) = \sigma((x,y),\ (u,v))$$

if and only if either $\text{Im}\, x\bar{u} = 0$ or $\text{Im}\, y\bar{v} = 0$. Hence S is not $\rho$-nonexpansive.

Observe finally that if K is a $\rho$-closed $\rho$-convex subset of B and $f: K \to B$ is $\rho$-nonexpansive, then the mapping $g: B \to B$ defined by $g(x) = f(R_K(x))$ is a $\rho$-nonexpansive extension of f to all of B. Therefore we shall assume from now on that the $\rho$-nonexpansive mappings under discussion are defined on the whole space B.

Remark. Let E be a subset of a Hilbert space H. According to the well-known Kirszbraun-Valentine theorem, any nonexpansive mapping $T:E \to H$ can be extended to a nonexpansive mapping $S:H \to H$. This is no longer true in the hyperbolic case. To see this, let B be the open unit ball of $C^2$,

$$E = \{(z,w) \in B: \text{either } z = 0 \text{ or } w = 0\},$$

and define a $\rho$-isometry $T:E \to E$ by $T(z,0) = (\bar{z},0)$ and $T(0,w) = (0,w)$. If there were a $\rho$-nonexpansive extension $S:B \to B$ of T, we would have

$$S(z/2,z/2) = S(\tfrac{1}{2}(z,0) \oplus \tfrac{1}{2}(0,z))$$
$$= \tfrac{1}{2}(\bar{z},0) \oplus \tfrac{1}{2}(0,z) = (\bar{z}/2,z/2)$$

and

$$S(\bar{z}/2,z/2) = (z/2,z/2)$$

for all $|z| < 1$. It would then follow that

$$\rho((w,w),(v,v)) = \rho((\bar{w},w),(\bar{v},v))$$

for all $|w|$, $|v| < \frac{1}{2}$. But this is clearly false. On the positive side, T. Kuczumow and A. Stachura have recently shown that an analog of the Kirszbraun-Valentine theorem does hold in the one-dimensional case (B=D) and in the real Hilbert ball (see Section 32).

## 23. FIXED POINTS

Let $T:B \to B$ be a $\rho$-nonexapnsive mapping. We shall call a sequence $\{y_n\} \subset B$ an approximating sequence for T if $\lim_{n\to\infty} \rho(y_n, Ty_n) = 0$. Using the properties of asymptotic centers and the method of proof of the analogous theorem in uniformly convex Banach spaces, we obtain the following basic result.

Theorem 23.1. Let $T: B \to B$ be a $\rho$-nonexpansive mapping. Then the following are equivalent:

(a) T has a fixed point;

(b) There exists a point x in B such that the sequence of iterates $\{T^n x\}$ is $\rho$-bounded;

(c) The sequence of iterates $\{T^n x\}$ is $\rho$-bounded for all x in B;

(d) There exists a $\rho$-bounded approximating sequence for T.

Proof. The asymptotic centers of the sequences in parts (b) and (d) are fixed points of T.

Other necessary and sufficient conditions for the existence of fixed points of $\rho$-nonexpanxive mappings will be discussed later.

As in the case of uniformly convex Banach spaces, we have the following result on the structure of the fixed point set F(T) of T.

Theorem 23.2. The fixed point set of a $\rho$-nonexpansive mapping $T : B \to B$ is $\rho$-closed and $\rho$-convex.

Even more can be said if T is holomorphic.

Theorem 23.3. The fixed point set of a holomorphic mapping $T: B \to B$ is affine.

*Proof.* Let A be a maximal affine subset of F(T). Suppose that $z \in F(T)$ does not belong to A, and let $w \in A$. Then the metric segment $[w,z]$ is contained in F(T) by Theorem 23.2. Since T is holomorphic, it follows that the entire one-dimensional disc spanned by z and w is contained in F(T). Thus the affine set spanned by z and A is contained in F(T). This contradicts the maximality of A. Hence A = F(T) and the proof is complete.

## 24. APPROXIMATING CURVES

If T is ρ-nonexpansive and $0 \leq t < 1$, the mapping $g: B \to B$ defined by $g(x) = (1-t)0 \oplus tTx$ for x in B is a strict ρ-contraction with constant t. Therefore it has a unique fixed point which we shall denote by z(t). We shall call the continuous mapping $z: [0,1) \to B$ an approximating curve (of the first kind). Note that

$$\rho(z(t), Tz(t)) = (1-t)\ \rho(0,Tz(t))$$

and that $z(t) = s(t)\ Tz(t)$ for a certain function $s: [0,1) \to [0,1)$ such that $\lim_{t\to 1} s(t) = 1$.

Suppose that K = F(T) is not empty, and let $R_K$ be the nearest point projection of B onto K. Consider $x = R_K(0)$.

Assume that for some t, $r = |z(t)| > |x|$. Then $z(t) = R_{B(0\ r)}\ Tz(t)$ and

$$\rho(x,z(t)) \geq \rho(Tx,Tz(t)) = \rho(x,Tz(t)) > \rho(x,z(t)).$$

This contradiction shows that $|z(t)| \leq |x|$ for all t. Hence

$$\lim_{t\to 1} \rho(z(t), T(z(t)) = 0.$$

Now let $\{z(t_n)\}$ converge weakly to z as $n \to \infty$ and $t_n \to 1$. Then z is the asymptotic center of $\{z(t_n)\}$ which is an approximating sequence for T. Therefore z belongs to K and $|z| \geq |x|$. But we also have $|z| \leq |x|$, so that $|z| = |x|$ and $z = x$. Now we see that $|x| \leq \liminf_{n\to\infty} |z(t_n)| \leq \limsup_{n\to\infty} |z(t_n)| \leq |x|$. Thus

$$\lim |z(t_n)| = |x|$$

and $\{z(t_n)\}$ converges strongly (and in the ρ-topology) to x. It follows that the strong $\lim_{t\to 1} z(t) = x$.

More generally, for any point a in B and $0 \leq t < 1$, let $z_a(t)$ be the unique fixed point of the mapping $g_a$: B → B defined by $g_a(x) = (1-t)a \oplus tTx$ for x in B. Thus we have

$$z_a(t) = (1-t)a \oplus tTz_a(t), \tag{24.1}$$

or equivalently,

$$z_a(t) = M_a s(t) M_{-a}(Tz_a(t))$$

with an appropriately chosen s: [0,1) → [0,1). Set $w(t) = M_{-a}z_a(t)$. Then w(t) is an approximating curve originating at 0 for the mapping $M_{-a}TM_a$. The fixed point set of this mapping is the image of F(T) under $M_{-a}$. It follows that the strong

$$\lim_{t\to 1} z_a(t) = R_K(a).$$

Thus we have established the following result.

Theorem 24.1. Let T: B → B be a ρ-nonexpansive mapping with a nonempty fixed point set K. For any a in B define the approximating curve $z_a$ : [0,1) → B by (24.1). Then the strong

$$\lim_{t\to 1} z_a(t) = R_K(a),$$

where $R_K$ is the nearest point projection of B onto K.

In the special case when $T \in \mathrm{Hol}(B)$ and $T(0) = 0$, K is an affine set containing the origin and $\lim_{t \to 1} z_a(t) = P(a)$, where P is the orthogonal projection of H onto the subspace spanned by K.

We now present a theorem of the Leray-Schauder type.

Theorem 24.2. A $\rho$-nonexpansive mapping $T : B \to B$ has a fixed point if and only if there exists $0 < r < 1$ such that $Ty \neq my$ for all $|y| = r$ and $m > 1$.

Proof. Let z be a fixed point of T and let $r > |z|$. If $Ty = my$ for some y of norm r and $m > 1$, then $y = R_{B(0,r)}(Ty)$ and $\rho(z,y) < \rho(z,Ty) = \rho(Tz,Ty) \leq \rho(z,y)$, a contradiction. Conversely, if the Leray-Schauder condition is satisfied, then $|z(t)| < r$ for all t. Hence $\lim_{t \to 1} \rho(z(t), Tz(t)) = 0$ and T has a fixed point by Theorem 23.1.

Remark. An analog of Theorem 16.1 in Chapter 1 can also be established.

Let $T: B \to B$ be a $\rho$-nonexpansive mapping. Changing our point of view, we now define for each $0 \leq t < 1$ a mapping $F_t : B \to B$ by $F_t(x) = z_x(t)$. In other words,

$$F_t(x) = (1-t)x \oplus tTF_t(x) \text{ for all } x \text{ in } B. \tag{24.2}$$

These mappings have the following properties:

(a) Each $F_t$ is $\rho$-nonexpansive;

(b) $F_t$ and T have the same fixed points;

(c) If the fixed point set $K = F(T)$ of T is not empty, then $\lim_{t \to 1} F_t(x) = R_K(x)$ for each x in B.

(d) The function $\phi: [0,1] \to [0,\infty)$ defined by $\phi(s) = \rho((1-s)x \oplus sF_t(x), (1-s)y \oplus sF_t(y))$ is non-increasing for each pair of points x and y in B.

To prove part (d), let $z = (1-s)x \oplus sF_t(x)$ and $w = (1-s)y \oplus sF_t(y)$. Noting that there exists $0 \leq p \leq 1$ such that $F_t(x) =$

$(1-p)z \oplus pTF_t(x)$ and $F_t(y) = (1-p)w \oplus pTF_t(y)$, we obtain

$$\phi(1) = \rho(F_t(x), F_t(y))$$

$$= \rho(F_p(z), F_p(w)) \leq \rho(z,w) = \phi(s).$$

Since $\phi$ is convex, the result follows.

We shall say that a mapping $T: B \to B$ is firmly nonexpansive (of the first kind) if for each $x$ and $y$ in $B$, the function $\phi$: $[0,1] \to [0,\infty)$ defined by $\phi(s) = \rho((1-s)x \oplus sTx, (1-s)y \oplus sTy)$ is non-increasing.

The family $FN_1$ of such mappings contains all the mappings $F_t$ defined by (24.2), as well as all the nearest point projections $R_K$ (see Section 19). It is closed under pointwise convergence. In view of (b) above, to each $T$ in $N$ there corresponds a family of mappings in $FN_1$ with the same fixed point sets.

These facts remind us of the situation in Hilbert space. Recall, however, that a mapping $T$ in Hilbert space is firmly nonexpansive if and only if $2T-I$ is nonexpansive. This fact does not carry over to our setting. Indeed we have seen in Section 22 that the mapping $2R_K \ominus I$ may fail to be $\rho$-nonexpansive for some $K$.

## 25. INVARIANT DOMAINS

Let $T: B \to B$ be a $\rho$-nonexpansive mapping. If $a$ is a fixed point of $T$, then all the $\rho$-balls centered at $a$ are invariant under $T$. Recall that such a ball $B(a,r)$ is an ellipsoid

$$E(a,k) = \{x \in B: \phi_a(x) < k\}$$

where $\phi_a(x) = |1 - (x,a)|^2/(1 - |x|^2)$ and $k = (1 - |a|^2)/(1 - \tanh^2 r)$. Now assume that the norm closure of the fixed point set of $T$ intersects the boundary of $B$ at $b$. Let $\{a_n\} \subset F(T)$ converge to $b$. Since $\phi_{a_n}(Tx) \leq \phi_{a_n}(x)$ for all $x$ in $B$, we see that

$$\phi_b(Tx) \leq \phi_b(x)$$

for all x. This means that E(b,k) is invariant under T for all k > 0. We have thus proved the following result.

Theorem 25.1. Let $T:B \to B$ be a ρ-nonexpansive mapping with a nonempty fixed point set F(T). Then the ellipsoids E(a,k) are T invariant for any a belonging to the norm closure of F(T).

We now turn to the case when T is fixed point free. Consider the approximating curve $z(t) = F_t(0)$, and let $\{z(t_n)\}$ converge weakly to u as $n \to \infty$ and $t_n \to 1$. We already know that

$$\lim_{t\to 1} |z(t)| = 1.$$

If we prove that $|u| = 1$, it will follow that z(t) converges strongly to u, and so does Tz(t).

To this end, assume that $|u| < 1$. Then

$$\sigma(Tz(t_n), Tu) \geq \sigma(z(t_n), u)$$

and

$$1 \geq \frac{1 - |z(t_n)|^2/s_n^2}{1 - |z(t_n)|^2} \geq \frac{(1 - |u|^2)|1 - (z(t_n), Tu)/s_n|^2}{(1 - |Tu|^2)|1 - (z(t_n),u)|^2},$$

where $0 < s_n < 1$ and $s_n \to 1$. Letting $n \to \infty$, we obtain $\sigma(u,Tu) \geq 1$ and $u = Tu$, a contradiction.

Let $x \in B$ and denote $z(t_n)$ by $z_n$. Since $\sigma(Tz_n,Tx) \geq \sigma(z_n,x)$, we obtain

$$\frac{|1 - (Tz_n,Tx)|^2}{1 - |Tx|^2} \leq \frac{1 - |Tz_n|^2}{1 - |z_n|^2}\,\frac{|1 - (z_n,x)|^2}{1 - |x|^2} \leq \frac{|1 - (z_n,x)^2|}{1 - |x|^2}.$$

Hence $\phi_u(Tx) \leq \phi_u(x)$. This means that all the ellipsoids $E(u,k)$ are invariant under T. As a matter of fact, u is the only point on the boundary of B with this property. To see this, suppose v is another such point. Then for sufficiently large k, $E(u,k) \cap E(v,k)$ is a nonempty $\rho$-bounded subset of B which is invariant under T. Hence T cannot be fixed point free. It also follows that the strong $\lim_{t \to 1} z(t)$ exists. In fact, all the approximating curves $z_a(t)$ converge to the same "sink point" as $t \to 1$.

We can sum up this discussion in the following way.

Theorem 25.2. If a $\rho$-nonexpansive mapping $T: B \to B$ is fixed point free, then there exists a unique point $e = e(T)$ of norm one such that all the ellipsoids $E(e,k)$, $k > 0$, are invariant under T. All the approximating curves $z_a(t)$, $a \in B$, converge strongly to e.

In this connection we observe that $e(F_t) = e(T)$ for the mappings $F_t$ defined in Section 24.

We continue with a consequence of Theorem 25.2.

Corollary 25.3. Let $T: B \to B$ be $\rho$-nonexpansive. If there exists x in B such that $\{T^n x\}$ converges weakly to a point in B, then T has a fixed point.

Proof. Let $\{T^n x\}$ converge weakly to z in B. If T is fixed point free, then $\limsup_{n \to \infty} |T^n x| = 1$. Since $\{T^n x\} \subset E(e(T),k)$ for sufficiently large k, $(z,e(T)) = 1$. Hence $|z| = 1$ and we have reached a contradiction. Therefore T must have a fixed point.

Here is another consequence of Theorem 25.2.

Theorem 25.4. If a norm continuous mapping $T : \overline{B} \to \overline{B}$ is $\rho$-nonexpansive on B, then it has a fixed point in $\overline{B}$.

Proof. If T has no fixed point in B, then e(T) must be a fixed point of T.

Corollary 25.5. A holomorphic mapping $T: B \to B$ with a continuous extension to $\overline{B}$ has a fixed point in $\overline{B}$.

This result, which solves an open problem raised by Harris [48], is due to Goebel, Sekowski and Stachura [43].

In this connection recall that if H is infinite-dimensional, then there are fixed point free continuous mappings (even homeomorphisms) of $\overline{B}$ into itself.

Theorem 25.4 does not imply that if T has no fixed point in B, then e(T) is its unique fixed point. In fact, this is not true even for automorphisms of B: the Möbius transformation $M_a$, $a \neq 0$, has two fixed points, $a/|a|$ and $-a/|a|$. (In this case, $e(M_a) = a/|a|$.) Also in the setting of Theorem 25.4, the set of fixed points in $\overline{B}$ may be larger than the norm closure of the nonempty fixed point set of T in B. Consider, for example, $H = C^2$ and the mapping

$$T(z,w) = (z,w^2).$$

In this case

$$F(T) = \{(z,0) : |z| < 1\}$$

but (0,1) is also a fixed point of T.

## 26. LOCALIZATION OF FIXED POINTS

If $T: B \to B$ is $\rho$-nonexpansive and $z \in F(T)$, then

$$\rho(z,Tx) \leq \rho(z,x)$$

for all $x \in B$. In other words, z belongs to $\cap \{\overset{+}{Eq}(x, Tx) : x \in B\}$. Conversely, if w belongs to this intersection of half-spaces, then $\rho(w,Tw) \leq \rho(w,w) = 0$ and w is a fixed point of T. Hence

$$F(T) = \cap \{\overset{+}{Eq}(x,Tx) : x \in B\}.$$

Let $\overline{F(T)}$ and $\overline{Eq}^{+}(x,y)$ denote the norm closures of F(T) and $Eq^{+}(x,y)$ respectively. Note that $z \in B$ belongs to $Eq^{+}(x,y)$ if and only if $\phi_z(y) \leq \phi_z(x)$, and that $w \in \partial B$ belongs to $\overline{Eq}^{+}(x,y)$ if and only if $\phi_w(y) \leq \phi_w(x)$. Therefore Theorem 25.1 shows that each w in $\overline{F(T)}$ belongs to $\overline{E}_q^{+}(x,Tx)$ for all $x \in B$. Conversely, if $w \in \partial B$ belongs to $\cap\{\overline{Eq}^{+}(x,Tx) : x \in B\}$ and F(T) is not empty, then each ellipsoid E(w,k) is invariant under T and therefore contains a fixed point of T. It follows that $w \in \overline{F(T)}$. Hence

$$\overline{F(T)} = \cap \{\overline{Eq}^{+}(x,Tx) : x \in B\}.$$

The following theorem sums up this discussion and also deals with the case when T is fixed point free.

<u>Theorem 26.1.</u> Let F(T) be the fixed point set of a $\rho$-nonexpansive $T : B \to B$. Then

(a) $F(T) = \cap \{Eq^{+}(x,Tx) : x \in B\}$,

(b) $\overline{F(T)} = \cap \{\overline{Eq}^{+}(x,Tx) : x \in B\}$,

and

(c) $\overline{F(T)} \cap \partial B = \cap\{\overline{Eq}^{+}(x,Tx) \cap \partial B : x \in B\}$.

If $F(T) = \phi$, then

(d) $\cap \{Eq^{+}(x,Tx) : x \in B\} = \phi$

and

(e) $\cap \{\overline{Eq}^{+}(x,Tx) \cap \partial B : x \in B\} = \{e(T)\}$.

<u>Corollary 26.2.</u> If $T \in N(B)$ and there exist n points $\{x_k : 1 \leq k \leq n\}$ in B such that

$$\cap \{\overline{Eq}^{+}(x_k,Tx_k) \cap \partial B : 1 \leq k \leq n\} = \phi$$

then T has a fixed point and $\overline{F(T)} \cap \partial B = \phi$. If $T \in Hol(B)$, then F(T) is a singleton.

If T is firmly nonexpansive (of the first kind), then Theorem 26.1 can be strengthened. For example, if $x \in B$, $T \in FN_1$ and $z \in$

$F(T)$, then $\rho(z, (1-t)x \oplus tTx)$ is non-increasing on $[0,1]$. Thus $z \in A(x,Tx) = E_q^+(x, 2Tx \ominus x)$ As a matter of fact, in this case Theorem 26.1 remains true when $E_q^+(x,Tx)$ is replaced by $E_q^+(x, 2Tx \ominus x)$.

In the next section we introduce firmly nonexpansive mappings of the second kind.

## 27. FIRMLY NONEXPANSIVE MAPPINGS

Recall that a mapping $T:B \to B$ is said to be firmly nonexpansive (of the first kind) if for each x and y in B, the function $\phi: [0,1] \to [0,\infty)$ defined by $\phi(s) = \rho((1-s)x \oplus sTx, (1-s)y \oplus sTy)$ is non-increasing. The family $FN_1$ of such mappings contains all the mappings $F_t$ defined in Section 24.

In this section we introduce a second type of firmly nonexpansive mappings.

We shall say that a mapping $T: B \to B$ is firmly nonexpansive of the second kind if for each x and y in B, the function $\psi: [0,1] \to [0,\infty)$ defined by $\psi(s) = \rho((1-s)x + sTx, (1-s)y + sTy)$ is non-increasing. The family of all such mappings will be denoted by $FN_2$.

We show now that to each T in N we can associate a family of mappings $\{G_t : 0 \leq t < 1\} \subset FN_2$ with the same fixed point sets.

Indeed, for any point a in B and $0 \leq t < 1$, let $y_a(t)$ be the unique fixed point of the strict contraction $g_a : B \to B$ defined by $g_a(x) = (1-t)a + tTx$ for x in B. We call $y_a : [0,1) \to B$ an approximating curve of the second kind. We now define the mappings $G_t : B \to B$ by $G_t(x) = y_x(t)$. In other words,

$$G_t(x) = (1-t)x + tTG_t(x) \text{ for all } x \text{ in } B. \tag{27.1}$$

We first show that each $G_t$ is $\rho$-nonexpansive.

<u>Lemma 27.1.</u> Each $G_t$ belongs to N.

Proof. Let $0 \leq t < 1$ and fix a point $w$ in $B$. Define a sequence of mappings $f_n : B \to B$ by $f_1(x) = (1-t)x + tTw$, $f_{n+1}(x) = (1-t)x + tT(f_n(x))$, $n \geq 1$. For each fixed $x \in B$, $G_t(x) = \lim_{n\to\infty} f_n(x)$ by Banach's fixed point theorem. Since

$$\rho(f_{n+1}(x), f_{n+1}(y)) \leq \max\{\rho(x,y), \rho(f_n(x), f_n(y))\}$$

for all $x$ and $y$ in $B$, it follows that

$$\rho(f_n(x), f_n(y)) \leq \rho(x,y),$$

and so $G_t$ is indeed $\rho$-nonexpansive. Now we are ready to prove that each $G_t$ belongs in fact to $FN_2$.

Theorem 27.2. Let $T: B \to B$ be $\rho$-nonexpansive. For each $0 \leq t < 1$, define $G_t : B \to B$ by (27.1). Then each $G_t$ is firmly nonexpansive of the second kind.

Proof. Denote $G_t = G(t,T)$ by $G$. We already know that $G$ is $\rho$-nonexpansive. Now let $0 \leq r < s \leq 1$, $u = (1-r)x + rGx$, $v = (1-s)x + sGx$, $w = (1-r)y + rGy$, and $z = (1-s)y + sGy$. We have to show that $\rho(v,z) \leq \rho(u,w)$. If $s = 1$, then $v = Gx = G(p,T)u$ and $z = Gy = G(p,T)w$, where $p = p(r) = t(1-r)/(1-tr)$. Therefore $\rho(v,z) = \rho(G(p,T)u, G(p,T)w) \leq \rho(u,w)$, as required. If $s < 1$, then $Gx = G(p(s), T)v$ and $Gy = G(p(s), T)z$. We also have $v = (1-q)u + qGx$ and $z = (1-q)w + qGy$, with $q = (s-r)/(1-r)$. Hence $v = G(q,G(p(s), T))u$ and $z = G(q, G(p(s), T))w$, so that once again $\rho(v,z) \leq \rho(u,w)$.

When $T$ is holomorphic, then all the mappings $\{f_n\}$ defined in the course of the proof of Lemma 27.1 are holomorphic. Since the sequence $\{f_n\}$ is uniformly bounded, $G_t$ is seen to be holomorphic too [52, p. 113]. Since it is also firmly nonexpansive of the second kind, we say it is firmly holomorphic. This class of mappings was first introduced in [42].

If $T : B \to B$ is $\rho$-nonexpansive and fixed point free, then so is $G_t$ for each $0 \leq t < 1$, and $e(G_t) = e(T)$. We now show that the approximating curves $y_a(t)$ also converge strongly to $e(T)$.

Theorem 27.3. Let $T: B \to B$ be $\rho$-nonexpansive. For each $0 \leq t < 1$ define the mapping $G_t : B \to B$ by (27.1). If $T$ is fixed point free, then the strong $\lim_{t\to 1} G_t(x) = e(T)$ for all $x$ in $B$.

Proof. Since $\lim_{t\to 1} (G_t(x) - TG_t(x)) = 0$ and $T$ is fixed point free, we must have $\lim_{t\to 1} |G_t(x)| = 1$. Consider an ellipsoid of the form $E(e,k)$. If $x$ belongs to $E(e,k)$, then so does $(1-t)x + tTz$ for all $z$ in $E(e,k)$ and $0 \leq t < 1$. Therefore all such ellipsoids are invariant under $G_t$. Hence $\phi_e(G_t(x)) \leq \phi_e(x)$, $\lim_{t\to 1} (G_t(x), e) = 1$, and the result follows.

We will discuss the behavior of $G_t$ when $F(T)$ is not empty in the next section.

In order to locate fixed points of mappings in $FN_2$, we consider now the set $B(x,y)$ consisting of all points $z \in B$ for which the function $\psi_z : [0,1) \to R^+$ defined by

$$\psi_z(t) = \rho(z, (1-t)x + ty)$$

is non-increasing. Equivalently, $z \in B(x,y)$ if

$$\tilde{\psi}_z(t) = \sigma(z, (1-t)x + ty)$$

is non-decreasing for $0 \leq t \leq 1$. A necessary condition for this to happen is that $\tilde{\psi}_z'(1) \geq 0$. A computation and some manipulations show that this latter condition is equivalent to

$$\mathrm{Re}(M_{-y}(z), Px + \sqrt{1 - |y|^2}\, Qx - y) \leq 0, \tag{27.2}$$

where $P = P_y$ is the orthogonal projection of $H$ onto the one-dimensional subspace spanned by $y$ and $Q = Q_y = I - P_y$. This condition can also be written in the following way:

$$\mathrm{Re}(M_{-y}(z),\ M_{-y}(x))(1 - (y,x)) \leq 0. \tag{27.3}$$

Let $S(x,y)$ be the set of all $z$ in $B$ that satisfy (27.3). This set is a closed half-space of the second kind which can be obtained by considering the sets $A((1-t)x + ty,y)$ (see Section 20) and letting $t \to 1-$.

Let $A$ be the complex line determined by $x$ and $y$, and let $a$ be the unique point of least norm in $A$. If $a = 0$, then $S(x,y)$ is a "cylinder" the intersection of which with $A$ is the one-dimensional half-space determined by (7.2). The discussion in Section 7 shows that in this case $S(x,y) = B(x,y)$. In the general case, the Möbius transformation $M_{-a}$ preserves Euclidean segments in $A$ and so we have $S(x,y) = M_a(S(M_{-a}x,\ M_{-a}y)) = M_a(B(M_{-a}x,\ M_{-a}y)) = B(x,y)$. In other words, (27.3) is not only a necessary condition for $z$ to belong to $B(x,y)$, but also a sufficient one. Thus

$$B(x,y) = \{z \in B\colon \mathrm{Re}(M_{-y}(z),\ M_{-y}(x))(1-(y,x)) \leq 0\}.$$

If $T \in FN_2$ and $z \in F(T)$, then $z \in B(x,Tx)$ for all $x \in B$. Hence $F(T) = \bigcap \{B(x,Tx)\colon\ x \in B\}$. As we have seen, each $B(x,Tx)$ is a closed half-space of the second kind.

## 28. NONEXPANSIVE RETRACTIONS

A nonexpansive retraction of $B$ onto $K \subset B$ is a $\rho$-nonexpansive mapping $Q : B \to K$ the fixed point set of which coincides with $K$. A subset $K$ of $B$ is said to be a nonexpansive retract of $B$ if there exists a nonexpansive retraction of $B$ onto $K$. Since the fixed point set of any $\rho$-nonexpansive mappings is $\rho$-closed and $\rho$-convex (Theorem 23.2), and since the nearest point projection onto any $\rho$-closed $\rho$-convex subset of $B$ is $\rho$-nonexpansive (Theorem 19.2), a subset $K \subset B$ is a nonexpansive retract of $B$ if and only if it is $\rho$-closed and $\rho$-convex. It is also clear that $K \subset B$ is a holomorphic retract of $B$ if and only if it is an affine set.

There may be however more than one nonexpansive retraction onto a given set. We have already seen in Section 6 that in the one-dimensional case the mapping that assigns Rez to z is a nonexpansive retraction onto (-1,1) which is different from the nearest point projection onto this geodesic. Even in the case of affine sets, there may be more than one holomorphic retraction.

Example 28.1. Let $H = C^2$ and $K = \{(z,0) : |z| < 1\}$. The nearest point projection onto K is $R_K(z,w) = (z,0)$. Two other holomorphic retractions of B onto K are $T(z,w) = (z + w^2/2, 0)$ and $S(z,w) = (z + 1 - \sqrt{1 - w^2}, 0)$.

On the other hand, we do have the following uniqueness result.

Theorem 28.2. The nearest point projection $R_K$ of B onto a $\rho$-closed $\rho$-convex subset K of B is the unique retraction of B onto K which is firmly nonexpansive of the first kind.

Proof. We know that $R_K$ is indeed firmly nonexpansive of the first kind (see Section 19). Suppose now that $Q : B \to K$ is another retraction which belongs of $FN_1$. For each x in B and z in K, $z \in A(x,Qx) = \overset{+}{Eq}(x,2Qx \ominus x)$. Hence $K \subset \overset{+}{Eq}(x,2Qx \ominus x)$ and $Qx = R_K(x)$ by Theorem 20.3.

Given a $\rho$-nonexpansive mapping $T: D \to D$ with a fixed point we consider now the mappings $G_t : D \to D$ defined by (27.1). Since each $G_t$ belongs to $FN_2$, $\rho(z, (1-s)x + sG_tx)$ decreases on [0,1] for each x in B and z in K, the fixed point set of T. In other words, each such z belongs to $B(x,G_tx)$. In the one-dimensional case we are looking at, this implies that

$$|x - G_tx| \leq |x - z| \qquad (28.1)$$

(see Section 7). In other words,

$$|x - G_t x| \leq d(x,K), \tag{28.2}$$

where $d(x,K) = \inf\{|x-z| : z \in K\}$. Since $\lim_{t \to 1} (G_t x - TG_t x) = 0$ and the curve $\{G_t x : 0 \leq t < 1\}$ is ρ-bounded, each subsequential limit y of $\{G_t x\}$ as $t \to 1$ is a fixed point of T. The inequality (28.2) shows that

$$|x-y| = d(x,K). \tag{28.3}$$

Since K is ρ-convex, it is not difficult to see that there is at most one point y in K satisfying (28.3). It follows that for each x in B, $\lim_{t \to 1} G_t x = E_K x$ exists. We call $E_K$ the Euclidean nearest point projection on K. It follows that $E_K$ is (the unique) retraction of D onto K which is firmly nonexpansive of the second kind. Thus we have established the following rather unexpected facts.

Theorem 28.3. Let $T: D \to D$ be a ρ-nonexpansive mapping with a nonempty fixed point set K. For each $0 \leq t < 1$ define the mapping $G_t : D \to D$ by (27.1). Then $\lim_{t \to 1} G_t(x) = E_K(x)$ for all x in D, where $E_K : D \to K$ is the Euclidean nearest point projection of D onto K.

Corollary 28.4. The Euclidean nearest point projection onto a ρ-closed ρ-convex subset of the open unit disc is ρ-nonexpansive.

When $\dim(H) \geq 2$, the situation is quite different. Let A be an affine set in B. Then A can be represented as $(a+E) \cap B$, where a is the unique point of least norm in A and E is a closed subspace of H. In this case the (hyperbolic) nearest point projection onto A is given by

$$R_A(x) = a + \frac{1 - |a|^2}{1 - (Q_E x, a)} P_E(x),$$

where $x \in B$, $P_E : H \to E$ is the orthogonal projection onto E, and $Q_E = I - P_E$. Thus $R_A$ is different from the Euclidean nearest point projection (which is not always even defined) unless $a = 0$. Nevertheless, a direct computation shows that for $G_t = G(t, R_A)$, the strong $\lim_{t \to 1} G_t(x) = R_A(x)$ for all $x \in B$. This shows, in particular, that $R_A$ is firmly holomorphic in the sense of Section 27. We are led, then, to the following conjecture.

Conjecture 28.5. Let $T: B \to B$ be a $\rho$-nonexpansive mapping with a nonempty fixed point set K. For each $0 \leq t < 1$ define the mapping $G_t: B \to B$ by (27.1). Then the strong $\lim_{t \to 1} G_t(x) = Q_K(x)$ for all x in B, where $Q_K : B \to K$ is the unique retraction of B onto K which is firmly nonexpansive of the second kind.

The argument at the beginning of Section 24 shows that $\lim_{t \to 1} G_t(0)$ always exists. Also, Ascoli's theorem shows that if $\dim(H) < \infty$, then each sequence $\{G_{t_n}\}$ with $t_n \to 1$ has a subsequence which converges pointwise on B to a retraction $Q : B \to K$ which belongs to $FN_2$. But we do not know if Q is unique.

## 29. FIXED POINTS OF AUTOMORPHISMS

Any automorphism T of B is continuous on $\overline{B}$ and therefore has at least one fixed point in $\overline{B}$ by Corollary 25.5. We also know that if F(T) in B is nonempty, then it is an affine set (Theorem 23.3). If $F(T) = \phi$, then one of the fixed points of T on the boundary of B is the "sink point" e(T). In the one-dimensional case the equation $z = Tz$ leads to a quadratic equation, so that T cannot have more than two fixed points. It is interesting that the same result is true in all dimensions [49].

Theorem 29.1. If $T \in \mathrm{Aut}(B)$ and F(T) in B is empty, then T has either one or two fixed points on $\partial B$.

<u>Proof.</u> We already know that $e = e(T)$ is a fixed point of $T$. Suppose $b \neq e$ is another fixed point. Since $T$ transforms affine sets onto affine sets and fixes $e$ and $b$, the complex line $A$ determined by $e$ and $b$ is invariant under $T$. Since $B$ is homogeneous, we may assume without any loss of generality that the origin belongs to $A$. It is obvious that $\bar{A}$ does not contain any other fixed points of $T$ besides $e$ and $b$. Assume now that $v$ is another fixed point of $T$ on $\partial B$. Then $v = \xi e + u$ where $|\xi| \leq 1$ and $(u,e) = 0$. Recall now that $T$ is of the form $U \circ M_{-a}$, where $U$ is unitary and $a = T^{-1}(0)$. Since $A$ is invariant under $T$, $a$ belongs to $A$ so that $a = \alpha e$ for some $|\alpha| < 1$. Note that

$$e = Te = UM_{-a}(e) = \frac{1 - \alpha}{1 - \bar{\alpha}} Ue$$

It follows that $Uu$ is also orthogonal to $e$.

Now we have

$$v = Tv = UM_{-a}(\xi e + u)$$

$$= U\left( \frac{\xi - \alpha}{1 - \xi\bar{\alpha}} e + \frac{\sqrt{1 - |a|^2}}{1 - \xi\alpha} u \right)$$

$$= \frac{\xi - \alpha}{1 - \xi\bar{\alpha}} \frac{1 - \bar{\alpha}}{1 - \alpha} e + \frac{\sqrt{1 - |a|^2}}{1 - \xi\bar{\alpha}} Uu.$$

Hence

$$\xi = (v,e) = \frac{\xi - \alpha}{1 - \xi\bar{\alpha}} \frac{1 - \bar{\alpha}}{1 - \alpha}.$$

Thus we see that $\xi$ is a fixed point of the automorphism

$$\frac{1 - \bar{\alpha}}{1 - \alpha} m_{-a} .$$

But this automorphism has no fixed point in D. Hence

$$|\xi| = 1, \; u = 0,$$

and the proof is complete.

## 30. ITERATIONS

Let T: B → B be a ρ-nonexpansive mapping, and let x be a point in B. In this section we study the behavior of the sequence of iterates $\{T^n x\} = \{x_n\}$. It is clear that if T is a strict contraction ($\rho(Tx,Ty) \leq k\rho(x,y)$ for some $k < 1$), then $\{x_n\}$ converges to the unique fixed point of T. This is the case, for example, if T is holomorphic and maps B strictly inside itself (see Section 13). On the other hand, $\{x_n\}$ may diverge even if T is an automorphism of B. We do have, however, a positive result for contractive mappings on finite-dimensional balls. (Recall that a mapping T: B → B is said to be contractive if $\rho(Tx,Ty) < \rho(x,y)$ for all $x \neq y$.)

In order to obtain this result, we recall (Theorem 25.2) that if a ρ-nonexpansive mapping T: B → B is fixed point free, then there exists a unique point e = e(T) of norm one such that all the ellipsoids E(e,k), k > 0, are invariant under T. We also need the following lemma.

<u>Lemma 30.1.</u> If a ρ-nonexpansive T: B → B is fixed point free and $\lim_{n\to\infty} |T^n x| = 1$ for some x in B, then the strong $\lim_{n\to\infty} T^n y = e(T)$ for all y in B.

<u>Proof.</u> Denote $T^n x$ by $x_n$, and let

$$\phi_e(x) = |1 - (x,e)|^2/(1 - |x|^2).$$

Since $\phi_e(Tx) \leq \phi_e(x)$, $\phi(x_n) \leq \phi_e(x)$ for all n. Hence

$$\lim_{n\to\infty} (x_n,e) = 1 \text{ and } \lim_{n\to\infty} x_n = e.$$

Now consider $y_n = T^n y$. Since $\rho(x_n,y_n) \leq \rho(x,y)$ for all n, $\lim_{n\to\infty} (x_n,y_n) = 1$, $\lim_{n\to\infty} |x_n - y_n| = 0$, and the strong $\lim_{n\to\infty} y_n = e(T)$.

Theorem 30.2. If H is finite-dimensional and $T: B \to B$ is contractive, then

(a) If T has a fixed point z in B, then $\lim_{n\to\infty} T^n x = z$ for all x in B;

(b) If T is fixed point free, then $\lim_{n\to\infty} T^n x = e(T)$ for all x in B.

Proof. Part (a) is an immediate consequence of Theorem 1.3 in Chapter 1. The same theorem shows that if T is fixed point free, then $\lim_{n\to\infty} |T^n x| = 1$ for all x in B. Therefore Part (b) follows from Lemma 30.1.

We deduce from Theorem 30.2 a one-dimensional result which is not true in higher dimensions.

Corollary 30.3. If $f \in \text{Hol}(D)$ is not an automorphism, then $\{f^n(z)\}$ converges for all z in D.

Proof. The holomorphic mapping f is contractive by Theorem 2.2 of Chapter 1.

If $T \neq I$ is an automorphism of D, then either it has a unique fixed point $z_0$ in D, or it has one or two fixed points on $\partial D$. In the first case the iterates of T do not converge unless they start at $z_0$. In the second case, the iterates do converge to e(T) (see [23]).

If $\dim(H) > 1$ (finite or infinite) and $T \in \mathrm{Aut}(B)$ does not have a fixed point in B, then T has either one or two fixed points on $\partial B$ (Theorem 29.1). When T has two fixed points on $\partial B$, the complex line determined by these two points is invariant under T. Therefore $\{T^n x\}$ converges to e(T) for all x in B by the one-dimensional case and Lemma 30.1. A recent infinite-dimensional example of A. Stachura shows that this is no longer true if T has only one fixed point on $\partial B$. On the other hand, if H is finite-dimensional and $T: B \to B$ is any holomorphic fixed point free mapping, then $\lim_{n\to\infty} T^n x = e(T)$ for all x in B. See the paper by B.D. MacCluer [66].

Returning to the finite-dimensional case, we note that any mapping T in Hol(B) which is not contractive is an isometry on a certain disc. Therefore Theorem 30.2 show that if $T \in \mathrm{Hol}(B)$ does not act isometrically on any disc, then its iterates always converge.

We conclude this discussion of contractive mappings with the observation that part (a) of Theorem 30.2 is not true in the infinite-dimensional case. To see this let $H = \ell^2$, and let the sequence $\{a_j\}$ satisfy $0 < a_j < 1$ with $\prod_{j=1}^{\infty} a_j > 0$. The mapping $T: B \to B$ defined by $(Tz)_j = 0$ for $j = 1$ and $(Tz)_j = a_{j-1}z_{j-1}$ for $j \geq 2$ is holomorphic and contractive, but its iterates do not converge unless they start at the origin.

We now turn to firmly nonexpansive mappings of both the first and second kinds (see Section 27). We begin with a lemma.

<u>Lemma 30.4.</u> Let $\{x_n\}$ and $\{z_n\}$ be two sequences in B. Suppose that for some y in B, $\lim_{n\to\infty} \rho(x_n,y) = \lim_{n\to\infty} \rho(z_n,y) = d$. Then the following are equivalent.

(a) $\lim_{n\to\infty} \rho((x_n + z_n)/2,y) = d$;

(b) $\lim_{n\to\infty} \rho(\frac{1}{2} x_n \oplus \frac{1}{2} z_n, y) = d$;

(c) $\lim_{n\to\infty} \rho(x_n,z_n) = 0$.

Proof. It is obvious that (c) implies (a) and (b). Let $d_n = \text{maz}\{\rho(x_n,y),\ \rho(z_n,y)\}$. We have

$$\rho(\tfrac{1}{2}\, x_n \oplus \tfrac{1}{2}\, z_n, y) \leq (1 - \delta(d_n,\ \rho(x_n,z_n)/d_n))d_n.$$

If (b) holds, then this inequality leads to a contradiction when $n \to \infty$ unless $\lim_{n\to\infty} \rho(x_n,z_n) = 0$. Thus (b) $\Rightarrow$ (c). Finally, assume that (a) holds. Recall (Section 15) that the closed $\rho$-ball $\overline{B}(y,d)$ is an ellipsoid which consists of all x in B that satisfy

$$|Px - u|^2/b^2r^2 + |Qx|^2/b^2r \leq 1,$$

where $b = \tanh(d)$, $P = P_y$, $Q = I-P$, $r = (1 - |y|^2)/(1 - b^2|y|^2)$, and $u = (1 - b^2)y/(1 - b^2|y|^2)$. Hence

$$\lim_{n\to\infty} \{|Px_n - u + Pz_n - u|^2/4b^2r^2 + |Qx_n + Qz_n|^2/4b^2r\} = 1.$$

Using the parallelogram law, we now see that

$$\lim_{n\to\infty} \{|P(x_n - z_n)|^2/4b^2r^2 + |Q(x_n - z_n)|^2/4b^2r\} = 0.$$

Therefore $\lim_{n\to\infty} |x_n - z_n| = 0$ and (c) follows because $\{x_n\}$ and $\{z_n\}$ are $\rho$-bounded.

Theorem 30.5. Let $T: B \to B$ be firmly nonexpansive of the first or second kind. If T has a fixed point, then for each x in B, the sequence of iterates $\{T^n x\}$ converges weakly to a fixed point of T.

Proof. Let y be a fixed point of T, and let $x_n = T^n x$. Since T is $\rho$-nonexpansive, $\{\rho(x_n,y)\}$ is a decreasing sequence and

$$d = \lim_{n\to\infty} \rho(x_n,y) = \lim_{n\to\infty} \rho(Tx_n,y)$$

exists. If $T \in FN_1$, then the function

$$\rho((1-s)x \oplus sTx, y)$$

is non-increasing and therefore

$$\rho(Tx_n, y) \leq \rho(\tfrac{1}{2} x_n \oplus \tfrac{1}{2} Tx_n, y) \leq \rho(x_n, y)$$

for all n. Hence $\lim_{n\to\infty} \rho(\frac{1}{2} x_n \oplus \frac{1}{2} Tx_n, y) = d$ and $\lim_{n\to\infty} \rho(x_n, Tx_n) = 0$ by Lemma 30.4. If $T \in FN_2$, then $\rho((1-s)x + sTx, y)$ is non-increasing and $\rho(Tx_n, y) \leq \rho((x_n + Tx_n)/2, y) \leq \rho(x_n, y)$. Therefore Lemma 30.4 shows that $\lim_{n\to\infty} \rho(x_n, Tx_n) = 0$ in this case too. Let a subsequence $\{x_{n_k}\}$ of $\{x_n\}$ converge weakly to z. Then z is the asymptotic center of $\{x_{n_k}\}$ by Proposition 2.14 and a fixed point of T by the proof of Theorem 23.1. Hence $\lim_{k\to\infty} \rho(x_{n_k}, z) = \lim_{n\to\infty} \rho(x_n, z)$ and z must be the (unique) asymptotic center of the whole sequence $\{x_n\}$. Consequently, $\{x_n\}$ converges weakly to z, as asserted.

We do not know if the convergence established in Theorem 30.5 is actually strong. As mentioned in Chapter 1, this is not true in general in the Hilbert space case. It would also be of interest to determine all the ρ-nonexpansive self-mappings of B for which the conclusion of Theorem 30.5 holds. For a step in this direction, see the paper by S. Reich [89].

We conclude this section by presenting a result for fixed point free mappings. It was announced in [42] for the firmly holomorphic case.

We first note that the proof of the implication (a) ⇒ (c) of Lemma 30.4 can also be used to establish the following fact.

<u>Lemma 30.6.</u> Let the point e belong to the boundary of B, and let $\{x_n\}$ and $\{z_n\}$ be two sequences in B. Suppose that

$$\lim_{n\to\infty} \phi_e(x_n) = \lim_{n\to\infty} \phi_e(z_n) = \lim_{n\to\infty} \phi_e((x_n + z_n)/2).$$

Then $\lim_{n\to\infty} |x_n - z_n| = 0$.

Now let $T \in FN_2$ be fixed point free and let $e = e(T)$ be the point obtained in Theorem 25.2.

Lemma 30.7. If $T \in FN_2$ is fixed point free and $x \in B$, then the function $f(s) = \phi_e((1-s)x + sTx)$ is non-increasing for $0 \leq s \leq 1$.

Proof. Let $y(t)$, $0 \leq t < 1$, be defined by $y(t) = tTy(t)$. If $0 \leq s_1 \leq s_2 \leq 1$, then

$$\sigma((1-s_1)x + s_1Tx,\ (1-s_1)y(t) + s_1Ty(t))$$

$$\leq \sigma((1-s_2)x + s_2Tx,\ (1-s_2)y(t) + s_2Ty(t)).$$

Since $|(1-s_1)y(t) + s_1Ty(t)| \leq |(1-s_2)y(t) + s_2Ty(t)|$ for all $0 \leq t < 1$, and since the strong $\lim_{t\to 1} y(t) = e$ by Theorem 27.3, some manipulation leads to the conclusion that $f(s_1)/f(s_2) \geq 1$, as required.

Theorem 30.8. Let $T: B \to B$ be firmly nonexpansive of the second kind. If $T$ is fixed point free, then for each $x$ in $B$, the sequence of iterates $\{T^n x\}$ converges strongly to $e(T)$.

Proof. Let $x_n = T^n x$. Since $\phi_e(Tx) \leq \phi_e(x)$, the sequence $\{\phi_e(x_n)\}$ is decreasing and $a = \lim_{n\to\infty} \phi_e(x_n)$ exists. Clearly

$$\lim_{n\to\infty} \phi_e(Tx_n) = a$$

too. By Lemma 30.7 we also have

$$\phi_e(Tx_n) \leq \phi_e((x_n + Tx_n)/2) \leq \phi_e(x_n) \text{ for all } n.$$

Thus

$$\lim_{n\to\infty} \phi_e((x_n + Tx_n)/2) = a,$$

and $\lim_{n\to\infty} |x_n - Tx_n| = 0$ by Lemma 30.6. Since T does not have a fixed point, this implies, by Theorem 23.1, that $\lim_{n\to\infty} |x_n| = 1$. The result now follows from Lemma 30.1.

Remark. Theorem 30.8 is also valid for firmly nonexpansive mappings of the first kind.

In connection with the difference between Theorem 15.3 of Chapter 1 and Theorem 30.8, it may be of interest to mention an analogy with Brownian motion: In Euclidean space of dimension greater or equal to three, almost all Brownian paths wander out to infinity, but with no asymptotic direction. In hyperbolic space (or more generally, in any complete simply connected Riemannian manifold with curvatures bounded between two negative constants), almost all paths tend to limits on the boundary. See the papers by J.J. Prat [75] and by D. Sullivan [93].

## 31. EQUIVALENT AND NONEQUIVALENT DOMAINS

We have observed in Section 7 that the space $(D,\rho)$ has many isometric models. In this section we wish to discuss infinite-dimensional domains that are equivalent to B.

We say that a domain G is a complex Banach space is equivalent to B if there is a biholomorphic mapping Q of B onto G. In this case we may define a metric $\rho_G$ on G by

$$\rho_G(x,y) = \rho(Q^{-1}x, Q^{-1}y).$$

The space $(G,\rho_G)$ is then isometric to $(B,\rho)$. Hence it is homogeneous, has a system of geodesics, and has the same modulus of convexity as $(B,\rho)$.

The difference between the one-dimensional case and the multi-dimensional case is that there are relatively few domains which are equivalent to B.

We begin our discussion with the analog of the upper half-plane model. Let e be a point in H of norm 1, and let the Cayley transform $C: B \to H$ be defined by $w = Cz = i(e+z)/(1-\xi)$, where $\xi = (z,e)$. The mapping C transforms any point of the form $\xi e$ with $|\xi| < 1$ onto a point $\eta e$ with $\mathrm{Im}\,\eta > 0$. Therefore it is useful to represent z and w as $z = (\xi, z')$ and $w = (\eta, w')$ where

$$\xi = (z,e), \quad \eta = (w,e), \quad z' = z - \xi e, \quad \text{and } w' = w - \eta e.$$

A computation shows that for any z in B,

$$\mathrm{Im}\,\eta - |w'|^2 = (1-|z|^2)/|1-\xi|^2,$$

and that C is invertible with

$$z = C^{-1}(w) = 2w/(i+\eta) - e.$$

Thus C is a biholomorphic mapping of B onto the Siegel domain S defined by

$$S = \{w \in H: \ \mathrm{Im}\,\eta > |w'|^2\}.$$

S is norm unbounded in H and its boundary $\partial S$ consists of all points $w \in H$ for which $\mathrm{Im}\,\eta = |w'|^2$. The image of e under C is not defined, but we may formally denote it by $\infty$ and add it to $\partial S$.

We now mention two interesting subgroups of Aut(S). The first consists of all the "non-isotropic" dilations $D_t$ defined by

$$D_t(w) = (t^2\eta, tw'), \ 0 < t < \infty.$$

These automorphisms are the analogs of the Möbius transformations $M_{se} : B \to B$ with $-1 < s < 1$. In fact,

$$D_t = C\, M_{se} C^{-1}$$

for $s = (t^2-1)/(t^2+1)$. For $t \neq 1$, each $D_t$ fixes only 0 and $\infty$.

The second subgroup of Aut(S) consists of the analogs of the horizontal translations of the one-dimensional upper half-plane. These mappings are of the form

$$h_a(w) = (\eta + \alpha + 2i(w',a'),\ w' + a'),$$

where $a = (\alpha,a') \in \partial S$; that is, $\operatorname{Im} \alpha = |a'|^2$. This subgroup induces a binary operation on $\partial S$ $(a*b = h_a(b))$ which makes it into a group (the so-called Heisenberg group). If $a \neq 0$, $h_a$ fixes only $\infty$.

We turn now to an example of a simple domain which is not equivalent to B.

Consider the product $B \times B$ of two unit balls. If H is infinite-dimensional, then $B \times B$ may be viewed as a domain in the same space H. If $\dim(H) = n < \infty$, then $B \times B = B^2$ is a domain in $H \times H = H^2$, the Hilbert space of dimension 2n. In any case, it is not difficult to see that $B^2$ is homogeneous. It is the unit ball of $H^2$ equipped with the norm

$$|(x,y)| = \max\ \{|x|,|y|\}.$$

Since this unit ball is homogeneous, all Schwarz-Pick systems assign the same metric $\rho = \rho_{B^2}$ to it (see Section 12), and we have

$$\rho_{B^2}((x,y),(u,v)) = \operatorname{argtanh}\ (|M_{-x}(u),M_{-y}(v))|)$$

$$= \operatorname{argtanh}(\max\{|M_{-x}(u)|,\ |M_{-y}(v)|\})$$

$$= \max\ \{\rho_B(x,u),\ \rho_B(y,v)\}.$$

Now let $a \neq 0$ be a point in B, and consider the three points $(a,-a)$, $(a,a)$ and $(a,0)$ in $B^2$. Since

$$\rho((a,0), (a,-a)) = \rho((a,0),(a,a)) = \rho((a,-a),(a,a))/2,$$

$(a,0)$ is a metric midpoint between $(a,-a)$ and $(a,a)$. But

$$\rho((0,0),(a,-a)) = \rho((0,0),(a,0)) = \rho((0,0),(a,a)),$$

so that $(B^2, \rho_{B^2})$ is not uniformly convex. Hence it is not isometric to $(B, \rho_B)$. Consequently, in an infinite dimensional Hilbert space H, $B^2$ is not equivalent to B, and the product of two n-dimensional balls is not equivalent to the 2n-dimensional ball.

These facts are known [45], but the concept of uniform convexity provides us with a very simple proof.

As a matter of fact, we can formulate the following criterion

<u>Theorem 31.1.</u> Let G be a domain in a Banach space X. If there exist four points x,y,u and z in G for which

$$\rho(x,u) = \rho(y,u) = \rho(x,y)/2$$

and

$$\rho(z,u) > (1-\delta_B(\rho_G(x,y)/r))r,$$

where $\delta_B$ is the modulus of convexity of B, $\rho_G$ is a Schwarz-Pick pseudometric on G, and $r = \max\{\rho_G(z,x), \rho_G(z,y)\}$, then G is not equivalent to B.

Intuitively speaking, G is not equivalent to B if it contains a "bad configuration" of four points. It is difficult, however, to apply this theorem because in general $\rho_G$ is not known explicitly.

Many domains are not equivalent to B because they are highly nonhomogeneous. It is known [10,35], for example, that every automorphism of the unit balls of the spaces, $L^p[0,1]$ and $\ell^p_n$ with $1 \leq p < \infty$, $p \neq 2$, and $n > 1$ is the restriction of a linear (norm) isometry. It is also known [54] that two complex Banach spaces are isometric if and only if their open unit balls are equivalent. Consequently, the theory we have developed for the Hilbert ball is unlikely to be extended to other domains. Nevertheless, we are able to apply our methods to certain domains which are not equivalent to B. Here is an example. We denote $\rho_{B^2}$ by $\rho$.

<u>Theorem 31.2.</u> A $\rho$-nonexpansive mapping $T : B^2 \to B^2$ has a fixed point if and only if there is a point z in $B^2$ such that the sequence of iterates $\{T^n z\}$ is $\rho$-bounded.

<u>Proof.</u> Let $T^n z = (x_n, y_n)$, and let $r_1$ and $r_2$ be the asymptotic radii of $\{x_n\}$ and $\{y_n\}$ respectively. If $r_1 = r_2$, then $A(\{(x_n, y_n)\})$, the set of asymptotic centers of $\{T^n z\}$, consists of exactly one point. This point must be a fixed point of T. If $r_1 < r_2$, for example, then $A(\{(x_n, y_n)\})$ is the product of

$$\{x \in B : r(x, \{x_n\}) \leq r_2\}$$

and $A(\{y_n\})$. This set is invariant under T and isometric to a $\rho$-bounded, $\rho$-closed and $\rho$-convex subset of B. Therefore it contains a fixed point of T by Theorem 23.1.

This result can be extended to $B^n$, the product of n Hilbert balls. In this connection we also mention the following recent extension of Corollary 25.5, due to A. Stachura: Any holomorphic self-mapping of $B^n$ which can be extended continuously to the boundary of $B^n$ must have a fixed point in $\overline{B^n}$.

## 32. THE REAL HILBERT BALL

Let $\{e_j\}$ be an orthonormal basis in a complex Hilbert space H of dimension d. Let $\tilde{H}$ be a real Hilbert space of the same dimension d over the reals. $\tilde{H}$ can be identified with the subset of H consisting of all z in H with real coefficients with respect to $\{e_j\}$. (That is, all z in H for which $\text{Im}(z,e_j) = 0$ for all j.) The unit ball $\tilde{B}$ of $\tilde{H}$ can be identified with all such z with norm smaller than one. Thus we may think of $\tilde{B}$ as a subset of B and consider the metric space $(\tilde{B},\rho)$.

We first observe that $\tilde{B}$ is $\rho$-convex. Indeed it is the fixed point set of the $\rho$-nonexpansive mapping which assigns to $z = \{\xi_j\}$ the point $\{\text{Re}\ \xi_j\}$. It is also clear that $\tilde{B}$ is $\rho$-closed in B. All the Möbius transformations $M_a$ with a in $\tilde{B}$ map $\tilde{B}$ into itself. These automorphisms also map (real) affine sets in $\tilde{B}$ onto affine sets. In particular, since the geodesics passing through the origin are affine, all geodesics in $(\tilde{B},\rho)$ are linear segments. Thus $\rho$-convexity in $(\tilde{B},\rho)$ coincides with linear convexity. It is obvious that $(\tilde{B},\rho)$ is uniformly convex. Its modulus of convexity equals that of $(B,\rho)$. The family $\tilde{N}$ of $\rho$-nonexpansive self-mappings of $\tilde{B}$ coincides with the restrictions to $\tilde{B}$ of all those $\rho$-nonexpansive self-mappings of B which leave $\tilde{B}$ invariant. Thus the fixed point theorems established for mappings in N by using uniform convexity carry over to $\tilde{N}$. Observe, however, that in $\tilde{B}$ approximating curves of the first kind coincide with those of the second kind. The same is true for firmly nonexpansive mappings of the first and second kind.

We now use the real Hilbert ball $\tilde{B}$ to study the almost fixed point property (AFPP) in the hyperbolic case. We say that a $\rho$-convex subset of B is geodesically bounded if its intersection with any geodesic in B is $\rho$-bounded. Such a set may be $\rho$-unbounded.

Example 32.1. Let $\{e_n\}$ be the standard orthonormal basis in $\ell^2$, and let $\{p_n\}$ be a sequence of positive numbers that converges to 1. Consider the set

$$C = \{ \sum_{n=1}^{\infty} \alpha_n p_n e_n : \alpha_n \geq 0, \sum_{n=1}^{\infty} \alpha_n \leq 1 \}.$$

This set is closed and $\rho$-closed. Since it is a convex subset of $\tilde{B}$, it is also $\rho$-convex. Although it is $\rho$-unbounded, it does not intersect the boundary of B. Hence it is geodesically bounded.

In contrast with the Banach space case, the AFPP for $\rho$-nonexpansive mappings implies the FPP in the hyperbolic case.

Theorem 32.2. A $\rho$-closed $\rho$-convex subset C of B has the fixed point property (and the almost fixed point property) for $\rho$-nonexpansive mappings if and only if it is geodesically bounded.

Proof. If C contains half a geodesic $\Gamma^+$, then the composition of the nearest point projection on $\Gamma^+$ with an appropriate Möbius transformation produces a mapping $T : C \to C$ for which

$$\inf\{\rho(x,Tx) : x \in C\}$$

is positive. Conversely, if $T : C \to C$ is fixed point free, consider an approximating curve $z_a(t)$ which starts at $a \in C$. Since $\lim_{t\to 1} z_a(t) = e(T)$, the "sink point" of T, the geodesic joining e(T) and a lies in C. Thus C cannot be geodesically bounded.

In view of Example 32.1, Theorem 32.2 shows that there are $\rho$-unbounded, $\rho$-closed and $\rho$-convex subsets of B which have the fixed point property for $\rho$-nonexpansive mappings. Once again, this is in marked contrast with the Euclidean (Hilbert space) case.

# 3

# SPHERICAL GEOMETRY

## 1. THE HILBERT SPHERE

Motivated by the Riemann sphere, we study in this section the unit sphere $S = \{x \in H : |x| = 1\}$ of a real Hilbert space H.

Consider the function $d : S \times S \to R$ defined by

$$d(x,y) = \arccos\,(x,y). \tag{1.1}$$

It is clear that d is non-negative and symmetric, and that $d(x,y) = 0$ if and only if $x = y$. In order to prove the triangle inequality, let $P_z$ denote the orthogonal projection of H onto the line spanned by $z \in H$, and let $Q_z = I - P_z$. Then for any x,y and z in S, we have

$$(x,y) = (P_z x,\ P_z y) = (Q_z x,\ Q_z y)$$

$$\geq\ (P_z x,\ P_z y) - |Q_z x|\,|Q_z y|$$

$$= (x,z)(y,z) - (1 - (x,z)^2)^{1/2}\ (1 - (y,z)^2)^{1/2}$$

$$= \cos\,(\arccos\,(x,z) + \arccos\,(y,z)).$$

Hence $d(x,y) \leq\ d(x,z) + d(z,y)$, as required.

Thus d is indeed a metric on S. We can also write

$$d(x,y) = 2 \arcsin \frac{|x-y|}{2} = 2 \arctan \frac{|x-y|}{|x+y|} .$$

The metric space (S,d) is bounded and its diameter equals $\pi$. ($d(x,y) = \pi$ if and only if $x = -y$.) S is complete and the d metric is equivalent to the norm metric. Since $d(x,y) \leq d(u,v)$ if and only if $|x-y| \leq |u-v|$, a mapping $T : S \to S$ is d-nonexpansive if and only if it is norm nonexpansive.

We shall say that a point z in S is a metric midpoint of x and y $(x \neq y)$ if

$$d(x,z) = d(z,y) = \frac{1}{2} d(x,y). \tag{1.2}$$

If $x = -y$, then z satisfies (1.2) if and only if it is orthogonal to x. Now assume that $x \neq -y$, and let u be the orthogonal projection of z onto the two-dimensional subspace of H spanned by x and y. Then $u = \alpha x + \beta y$ and

$$\arccos(x,u) = \arccos(y,u) = \frac{1}{2} \arccos (x,y). \tag{1.3}$$

Hence $\alpha = \beta$, $(x,y) = 2(x,u)^2 - 1$, and $\alpha^2 = 1/2(1+(x,y))$.
Since

$$d(x,u) < \pi/2 ,$$

$$(x,u) > 0, \ \alpha = 1/|x+y|, \text{ and } u = \frac{x+y}{|x+y|} .$$

Since $|u| = 1$, z must equal u. Thus we see that any two points x and y in S with $x \neq -y$ have exactly one metric midpoint, namely

$$z = \frac{x+y}{|x+y|} . \tag{1.4}$$

It follows that any two points x and y in S with $x \neq -y$ can be joined by a unique metric segment which is isometric to the interval $[0,d(x,y)]$. Once again we denote this segment by $[x,y]$ and its midpoint by $\frac{1}{2}x \oplus \frac{1}{2}y$. We can represent this segment by

$$z(t) = \frac{(1-t)x+ty}{|(1-t)x + ty|}, \qquad 0 \leq t \leq 1,$$

but this is not an isometric representation. We shall call the intersection of S with a two-dimensional subspace of H a "great circle" or a geodesic. Using this terminology, we see that metric segments are arcs of "great circles". (If $x = -y$, there are infinitely many such segments joining x and y.)

Balls in S have a simple structure.

$$B(x,r) = \{y \in S: d(x,y) < r\}$$

$$= \{y \in S: \quad (x,y) > \cos r\} = \{y \in S: \quad |P_x y| > \cos r\}.$$

If $r < \pi/2$, we can also write

$$B(x,r) = \{y \in S: |x \cos r - y| < \sin r\}.$$

Any hyperplane in S cuts S into two balls and the closed ball $\overline{B}(x,r) = S \setminus B(-x, \pi-r)$.

## 2. UNIFORM CONVEXITY

As in the Banach space and hyperbolic cases, we define the modulus of convexity of $(S,d)$ by

$$\delta(r,\varepsilon) = \inf\{1 - \frac{1}{r} d(a, \frac{1}{2}x \oplus \frac{1}{2}y)\},$$

where the infimum is taken over all points a,x and y in S satisfying $d(a,x) \leq r$, $d(a,y) \leq r$, and $d(x,y) \geq \varepsilon r$. This time $0 < r < \pi$, and $0 \leq \varepsilon < 2$ if $r \leq \pi/2$, but $0 \leq \varepsilon < 2(\pi/r-1)$ for $r \geq \pi/2$.

Let $\dim(H) > 2$. In order to compute $\delta$, let $a,x,v$ in $S$ satisfy $d(a,x) \leqslant r$, $d(a,y) \leqslant r$, and $d(x,v) \geqslant \varepsilon r$. Then $(a,x) \geqslant \cos r$, $(a,y) \geqslant \cos r$, and $(x,y) \leqslant \cos(\varepsilon r)$. Hence

$$\left(a, \frac{1}{2}x \oplus \frac{1}{2}y\right) = \left(a, \frac{x+y}{|x+y|}\right) = \frac{(a,x) + (a,y)}{\sqrt{2(1+(x,y))}} \geqslant \frac{\cos r}{\cos(\varepsilon r/2)}\,.$$

Since this inequality is the best possible, we conclude that

$$\delta(r,\varepsilon) = 1 - \frac{1}{r}\arccos\left(\frac{\cos r}{\cos(\varepsilon r/2)}\right).$$

Note that for $\varepsilon > 0$, $\delta(r,\varepsilon)$ is positive only for $r < \pi/2$. $\delta(\pi/2,\varepsilon) = 0$ for all $\varepsilon$, and $\delta(r,\varepsilon)$ is negative for $r > \pi/2$. It is still true, however, that

$$\lim_{r\to 0} \delta(r,\varepsilon) = \delta_H(\varepsilon) = 1 - (1 - \varepsilon^2/4)^{1/2}\,.$$

Let us call a subset $C$ of $S$ d-convex if any two distinct points in $C$ can be joined by a unique metric segment contained in $C$. Observe that a d-convex set can contain two antipodal points.

In our present setting, the intersection of a decreasing sequence of nonempty closed d-convex sets may well be empty. To see this, let $H = \ell^2$,

$$S = \{x \in H:\ |x| = 1\},\ S^+ = \{x \in S:\ x_i \geqslant 0 \text{ for all } i\},$$

and $C_n = \{x \in S^+ :\ x_i = 0 \text{ for } 1 \leqslant i \leqslant n\}$. However, the usual proof does yield the following weaker analog of the intersection theorem. We shall say that the sequence $\{C_n\}$ is strictly contained in a hemisphere if there exist $x$ in $S$ and $r < \pi/2$ such that $C_n \subset B(x,r)$ for all $n$.

<u>Theorem 2.1.</u> Let $\{C_n\colon n=1,2,\ldots\}$ be a decreasing sequence of nonempty closed d-convex subsets of $S$. If the sequence $\{C_n\}$ is

strictly contained in a hemisphere, then the intersection $\cap\{C_n : n=1,2,...\}$ is a nonempty closed d-convex subset of S.

Note that the intersection of d-convex sets is always d-convex, as is the union of a linearly ordered (by inclusion) family of such sets. On the other hand, the closure of the d-convex ball $B(x,\pi/2)$ is not d-convex.

Let C be a d-convex subset of (S,d). For x and y in C, let $(1-t)x \oplus ty$ denote the unique point z on the metric segment [x,y] satisfying $d(x,z) = td(x,y)$ and $d(z,y) = (1-t)d(x,y)$. We shall say that a function $f : C \to (-\infty,\infty)$ is quasi-d-convex if

$$f((1-t)x \oplus ty) \leq \max\{f(x),f(y)\} \qquad (2.1)$$

for all x and y in K and all $0 \leq t \leq 1$.

The following result is a consequence of Theorem 2.1.

Proposition 2.2. Let C be a closed, d-convex subset of (S,d), and let $f: C \to [0,\infty)$ be a quasi-d-convex function. If f is continuous and C is strictly contained in a hemisphere, then f attains its minimum on C. If, in addition,

$$f\left(\frac{1}{2}x \oplus \frac{1}{2}y\right) < \max\{f(x),f(y)\}$$

for all $x \neq y$, then it attains its minimum at exactly one point.

We continue with an application of Proposition 2.2. We denote $\inf\{d(x,z): z \in C\}$ by $d(x,C)$.

Theorem 2.3. Let C be a closed d-convex subset of S. If $x \in S$ and $d(x,C) < \pi/2$, then there exists exactly one point y in C such that $d(x,y) = d(x,C)$.

Proof. Choose a number $d(x,C) < p < \pi/2$, and apply Proposition 2.2 to the set $D = \{z \in C: d(x,z) \leq p\}$ and the function $f: D \to [0,p]$ defined by $f(z) = d(x,z)$.

In the present setting the nearest point projection is not defined on all of S and is not nonexpansive in general.

Finally, we note that for any fixed $a \in S$ and $0 \leq t < 1$, the mapping $f(x) = (1-t)a \oplus tx$ is well defined for all x in S except $x = -a$. Although f is not nonexpansive in general, it is a strict contraction when restricted to the ball $B(a,\pi/2)$. Even in this case, however, its Lipschitz constant is greater than t.

## 3. ASYMPTOTIC CENTERS

Let $\{x_n\}$ be a sequence in S, and consider the functional $f : S \to [0,\pi]$ defined by

$$f(x) = \limsup_{n\to\infty} d(x_n,x).$$

As before, the asymptotic radius of $\{x_n\}$ (with respect to S) is defined by

$$r(\{x_n\}) = \inf\{f(x) : x \in S\},$$

and a point z in S is said to be an asymptotic center of the sequence $\{x_n\}$ if

$$f(z) = r(\{x_n\}).$$

In contrast with previous cases, asymptotic centers in S are not unique. We do have, however, the following result.

Theorem 3.1. Let $\{x_n\}$ be a sequence in S. If the asymptotic radius $r(\{x_n\}) < \pi/2$, then $\{x_n\}$ has exactly one asymptotic center.

Proof. Denote $r(\{x_n\})$ by r, let $\{p_k\}$ be a positive sequence that decreases to 0, and let $f(x) = \limsup_{n\to\infty} d(x_n,x)$. For each $k \geq 1$, let $b_k$ be the diameter of the set

$$D_k = \{x \in S : f(x) \leq r + p_k\}.$$

Assume that $b = \lim_{k\to\infty} b_k$ (hence $r$) are positive. For sufficiently large $k$, let $x$ and $y$ be two points in $D_k$ such that $d(x,y) \geq b_k - p_k$. There is an $m$ such that $d(x_m,x) \leq r + 2p_k < \pi/2$, $d(x_m,y) \leq r + 2p_k$, and $d(x_m, \frac{1}{2}x \oplus \frac{1}{2}y) \geq r - p_k$. Set

$$\varepsilon_k = (b_k - p_k)/(r + 2p_k).$$

We have then

$$r - p_k \leq d(x_m, \frac{1}{2}x \oplus \frac{1}{2}y) \leq (1-\delta(r+2p_k, \varepsilon_k))(r+2p_k).$$

Letting $k \to \infty$, we obtain the contradiction

$$r \leq (1 - \delta(r,b/r))r.$$

Therefore $b = 0$ and the result follows.

## 4. FIXED POINTS

Let $T : S \to S$ be a nonexpansive mapping. For $x$ in $S$ consider the sequence of iterates $\{T^n x\}$. If for some $x \in S$ the asymptotic radius $r(\{T^n x\}) < \pi/2$, then Theorem 3.1 provides us with a unique asymptotic center for $\{T^n x\}$. Denote this asymptotic center by $A(\{T^n x\})$. Since $\limsup_{n\to\infty} d(T^n x,Tz) \leq \limsup_{n\to\infty} d(T^n x,z)$ for all $z \in S$, we see that $A(\{T^n x\})$ is a fixed point of $T$. Similarly, if there is a sequence $\{y_n\} \subset S$ such that $d(y_n, Ty_n) \to 0$ and $r(\{y_n\}) < \pi/2$, then the asymptotic center $A(\{y_n\})$ of such an approximating sequence is again a fixed point of $T$.

We have thus established the following fixed point theorem.

Theorem 4.1. Let $T : S \to S$ be a nonexpansive mapping. Then the following are equivalent:

(a) T has a fixed point;

(b) There exists a point x in S such that $r(\{T^n x\}) < \pi/2$;

(c) There exists an approximating sequence $\{y_n\}$ for T such that the asymptotic radius $r(\{y_n\}) < \pi/2$.

We also have the following fact.

Proposition 4.2. Let F(T) be the fixed point set of a nonexpansive $T : S \to S$. Then for any $a \in S$ and $r < \pi/2, F(T) \cap \overline{B}(a,r)$ is closed and d-convex.

Theorem 4.1 is of course an analog of the corresponding theorems in the Banach space and hyperbolic cases. As a matter of fact, it can be extended to certain other metric spaces. Let us say that a metric space (M,D) is locally uniformly convex if to each point a in M there correspond a number R = R(a) and a function $\delta_a : (0,R) \times (0,2] \to (0,1]$ such that

(a) Each two points in the ball B(a,R) can be joined by a unique metric segment which is isometric to the real interval [0, D(x,y)]; and

(b) For any $0 < r < R$ and $0 < \varepsilon \leq 2$,

$$\left.\begin{array}{l} D(a,x) \leq r \\ D(a,y) \leq r \\ D(x,y) \geq \varepsilon r \end{array}\right\} \Rightarrow D(a, \tfrac{1}{2} x \oplus \tfrac{1}{2} y) \leq (1 - \delta_a(r,\varepsilon))r .$$

If $R(a) = \infty$ for all $a \in M$ and $\delta(r,\varepsilon) = \inf \{\delta_a(r,\varepsilon) : a \in M\}$ is positive for all $0 < r < \infty$ and $0 < \varepsilon \leq 2$, we shall say that M is uniformly convex.

Using this terminology, $(B,\rho)$ is seen to be uniformly convex, while (S,d) is locally uniformly convex.

Now let (M,D) be locally uniformly convex and let $T : M \to M$ be nonexpansive. To see how Theorem 4.1 may be extended, consider a sequence of iterates $\{T^n x\}$ and its asymptotic radius $r(\{T^n x\})$. Suppose that

$$R = \liminf_{n\to\infty} R(T^n x) > 0$$

and that

$$\liminf_{n\to\infty} \delta_{T^n x}(r,\varepsilon) > 0$$

for all $0 < r < R$ and $0 < \varepsilon \leq 2$. Then the proofs of Theorem 3.1 and 4.1 show that if $r(\{T^n x\}) < R$, then the unique asymptotic center of $\{T^n x\}$ must be a fixed point of T. Part (c) of Theorem 4.1 can be generalized in a similar manner.

# REFERENCES


1. D.E. Alspach, A fixed point free nonexpansive map, Proc. Amer. Math. Soc. 82(1981), 423-424.
2. J.B. Baillon, Un théorème de type ergodique pour les contractions nonlinéaires dans un espace de Hilbert, C.R. Acad. Sci. Paris 280(1975), 1511-1514.
3. J.B. Baillon, Comportement asymptotique des itérés de contractions nonlinéaires dans les espaces $L^p$, C.R. Acad. Sci. Paris 286(1978), 157-159.
4. J.B. Baillon, Quelques aspects de la théorie des points fixes dans les espaces de Banach, preprint.
5. J.B. Baillon, R.E. Bruck, and S. Reich, On the asymptotic behavior of nonexpansive mappings and semigroups in Banach spaces, Houston J. Math. 4(1978), 1-9.
6. J.B. Baillon and W.O. Ray, Fixed points and approximate fixed points of nonexpansive mappings, preprint.
7. J.B. Baillon and R. Schöneberg, Asymptotic normal structure and fixed points of nonexpansive mappings, Proc. Amer. Math. Soc. 81(1981), 257-264.
8. L.P. Belluce, W.A. Kirk, and E.F. Steiner, Normal structure in Banach spaces, Pacific J. Math. 26(1968), 433-440.
9. Y. Benyamini and Y. Sternfeld, Spheres in infinite dimensional normed spaces are Lipschitz contractible, Proc. Amer. Math. Soc. 88(1983), 439-445.
10. R. Braun, W. Kaup, and H. Upmeier, On the automorphisms of circular and Reinhardt domains in complex Banach spaces, Manuscripta Math. 25(1978), 97-133.
11. H. Brezis, "Opérateurs Maximaux Monotones", North Holland, Amsterdam, 1973.

12. M.S. Brodskii and D.P. Milman, On the center of a convex set, Dokl. Akad. Nauk SSSR 59(1948), 837-840.

13. F.E. Browder, Nonexpansive nonlinear operators in a Banach space, Proc. Nat. Acad. Sci. U.S.A. 54(1965), 1041-1044.

14. F.E. Browder, Semicontractive and semiaccretive nonlinear mappings in Banach spaces, Bull. Amer. Math. Soc. 74(1968), 660-665.

15. F.E. Browder, "Nonlinear Operators and Nonlinear Equations of Evolution in Banach Spaces", Amer. Math. Soc., Providence, R.I., 1976.

16. A. L. Brown, A rotund reflexive space having a subspace of codimension two with a discontinuous metric projection, Michigan Math. J. 21(1974), 145-151.

17. R. E. Bruck, Properties of fixed point sets of nonexpansive mappings in Banach spaces, Trans. Amer. Math. Soc. 179(1973), 251-262.

18. R.E. Bruck, Nonexpansive projections on subsets of Banach spaces, Pacific J. Math. 47(1973), 341-355.

19. R. E. Bruck, On the almost convergence of iterates of a non-expansive mapping in Hilbert space and the structure of the weak omega-limit set, Israel J. Math. 29(1978), 1-16.

20. R.E. Bruck, A simple proof of the mean ergodic theorem for nonlinear contractions in Banach spaces, Israel J. Math. 32(1979), 107-116.

21. R.E. Bruck and S. Reich, Nonexpansive projections and resolvents of accretive operators in Banach spaces, Houston J. Math. 3(1977), 459-470.

22. R.E. Bruck and S. Reich, Accretive operators, Banach limits, and dual ergodic theorems, Bull. Acad. Polon. Sci. 29(1981), 585-589.

23. R.B. Burckel, Iterating analytic self-maps of discs, Amer. Math. Monthly 88(1981), 396-407.

24. W. L. Bynum, A class of spaces lacking normal structure, Compositio Math. 25(1972), 233-236.

25. J. Caristi, Fixed point theorems for mappings satisfying inwardness conditions, Trans. Amer. Math. Soc. 215(1976), 241-251.

26. J.A. Clarkson, Uniformly convex spaces, Trans. Amer. Math. Soc. 40(1936), 396-414.

27. M.M. Day, Reflexive Banach spaces not isomorphic to uniformly convex spaces, Bull. Amer. Math. Soc. 47(1941), 313-317.

28. D.J. Downing and B. Turett, Some properties of the characteristic of convexity relating to fixed point theory, preprint.

29. C.J. Earle and R.S. Hamilton, A fixed point theorem for holomorphic mappings, Proc. Symp. Pure Math., Vol. 16, Amer. Math. Soc., Providence, R.I., 1970, pp. 61-65.

30. M. Edelstein, On fixed and periodic points under contractive mappings, J. London Math. Soc. 37(1962), 74-79.

31. M. Edelstein, On nonexpansive mappings of Banach spaces, Proc. Camb. Phil. Soc. 60(1964), 439-447.

32. I. Ekeland, On the variational principle, J. Math. Anal. Appl. 47(1974), 324-353.

33. P. Enflo, Banach spaces which can be given an equivalent uniformly convex norm, Israel J. Math. 13(1972), 281-288.

34. D.G. deFigueiredo and L.A. Karlovitz, On the radial projection in normed spaces, Bull. Amer. Math. Soc. 73(1967), 364-368.

35. R.J. Fleming and J.E. Jamison, Some Banach spaces on which all biholomorphic automorphisms are linear, J. Math. Anal. Appl. 87(1982), 127-133.

36. T. Franzoni and E. Vesentini, "Holomorphic Maps and Invariant Distances", North Holland, Amsterdam, 1980.

37. A. Genel and J. Lindenstrauss, An example concerning fixed points, Israel J. Math. 22(1975), 81-86.

38. K. Goebel, Fixed points and invariant domains of holomorphic mappings of the Hilbert ball, Nonlinear Analysis 6(1982), 1327-1334.

39. K. Goebel, W.A. Kirk, and R.L. Thele, Uniformly Lipschitzian families of transformations in Banach spaces, Canad. J. Math. 26(1974), 1245-1256.

40. K. Goebel and T. Kuczumow, A contribution to the theory of nonexpansive mappings, Bull. Cal. Math. Soc. 70(1978), 355-357.

41. K. Goebel and T. Kuczumow, Irregular convex sets with the fixed point property for nonexpansive mappings, Colloq. Math. 40(1979), 259-264.

42. K. Goebel and S. Reich, Iterating holomorphic self-mappings of the Hilbert Ball, Proc. Japan Acad. 58(1982), 349-352.

43. K. Goebel, T. Sekowski, and A. Stachura, Uniform convexity of the hyperbolic metric and fixed points of holomorphic mappings in the Hilbert ball, Nonlinear Analysis 4(1980), 1011-1021.

44. D. Göhde, Zum Prinzip der kontraktiven Abbildung, Math. Nachr. 30(1965), 251-258.

45. S.J. Greenfield and N.R. Wallach, The Hilbert ball and bi-ball are holomorphically inequivalent, Bull Amer. Math. Soc. 77(1971), 261-263.

46. B. Halpern, Fixed points of nonexpanding maps, Bull. Amer. Math. Soc. 73(1967), 957-961.

47. O. Hanner, On the uniform convexity of $L^p$ and $\ell^p$, Ark. Mat. 3(1956), 239-244.

48. L.A. Harris, Schwarz-Pick systems of pseudometrics for domains in normed linear spaces, "Advances in Holomorphy", North Holland, Amsterdam, 1979, pp. 345-406.

49. T.L. Hayden and T.J. Suffridge, Biholomorphic maps in Hilbert space have a fixed point, Pacific J. Math. 38(1971), 419-422.

50. R. Haydon, E. Odell, and Y. Sternfeld, A fixed point theorem for a class of star-shaped sets in $c_0$, Israel J. Math. 38(1981), 75-81.

51. J.W. Helton, Non-Euclidean functional analysis and electronics, Bull. Amer. Math. Soc. 7(1982), 1-64.

52. E. Hille and R.S. Phillips, "Functional Analysis and Semi-groups", Amer. Math. Soc., Providence, R.I., 1957.

53. L.A. Karlovitz, On nonexpansive mappings, Proc. Amer. Math. Soc. 55(1976), 321-325.

54. W. Kaup and H. Upmeier, Banach spaces with biholomorphically equivalent unit balls are isomorphic, Proc. Amer. Math. Soc. 58(1976), 129-133.

55. B. Kawohl and R. Ruhl, Periodic solutions of nonlinear heat equations under discontinuous boundary conditions, preprint.

56. W.A. Kirk, A fixed point theorem for mappings which do not increase distances, Amer. Math. Monthly 72(1965), 1004-1006.

57. W.A. Kirk, Fixed point theory for nonexpansive mappings, Lecture Notes in Math., Vol. 886, Springer, Berlin and New York, 1981, pp. 484-505.

58. W.A. Kirk, Fixed point theory for nonexpansive mappings II, Contemporary Math. 18(1983), 121-140.

59. W.A. Kirk and Y. Sternfeld, The fixed point property for non-expansive mappings in certain product spaces, preprint.

60. E. Kohlberg and A. Neyman, Asymptotic behavior of nonexpansive mappings in normed linear spaces, Israel J. Math. 38(1981), 269-275.

61. B.R. Kripke, Unpublished example, 1971.

62. E.A. Lifschitz, Fixed point theorems for operators in strongly convex spaces, Voronez. Gos. Univ. Trudy Mat. Fak. 16(1975), 23-28.

63. T.C. Lim, Asymptotic centers and nonexpansive mappings in some conjugate spaces, Pacific J. Math. 90(1980), 135-143.

64. T.C. Lim, Fixed point theorems for uniformly Lipschitzian mappings in $L^p$ spaces I,II, preprint.

65. G.G. Lorentz, A contribution to the theory of divergent series, Acta Math. 80(1948), 167-190.

66. B.D. MacCluer, Iterates of holomorphic self-maps of the unit ball in $C^N$, Michigan Math. J. 30(1983), 97-106.

67. B. Maurey, Points fixes des contractions de certains faiblement compacts de $L^1$, preprint.

68. D.P. Milman, On some criteria for the regularity of spaces of type (B), Dokl. Akad. Nauk SSSR 20(1938), 243-246.

69. J. Milnor, Hyperbolic geometry: the first 150 years, Bull. Amer. Math. Soc. 6(1982), 9-24.

70. G. Nördlander, The modulus of convexity in normed linear spaces, Ark. Mat. 4(1960), 15-17.

71. B. Nowak, On the lipschitzian retraction of the unit ball in infinite-dimensional Banach spaces onto its boundary, Bull. Acad. Polon. Sci. 27(1979), 861-864.

72. Z. Opial, Weak convergence of the sequence of successive approximations for nonexpansive mappings, Bull. Amer. Math. Soc. 73(1967), 591-597.

73. B.J. Pettis, A proof that every uniformly convex space is reflexive, Duke Math. J. 5(1939), 249-253.

74. A.T. Plant and S. Reich, The asymptotics of nonexpansive iterations, J. Functional Analysis, to appear.

75. J.J. Prat, Étude asymptotique et convergence angulaire du mouvement brownien sur une variété à courbure négative, C.R. Acad. Sci. Paris 280(1975), 1539-1542.

76. W.O. Ray, Nonexpansive mappings on unbounded convex domains, Bull. Acad. Polon. Sci. 26(1978), 241-245.

77. W.O. Ray, The fixed point property and unbounded sets in Hilbert space, Trans. Amer. Math. Soc. 258(1980), 531-537.

78. S. Reich, Fixed points of condensing functions, J. Math. Anal. Appl. 41(1973), 460-467.

79. S. Reich, On fixed point theorems obtained from existence theorems for differential equations, J. Math. Anal. Appl. 54(1976), 26-36.

80. S. Reich, The fixed point property for nonexpansive mappings I, II, Amer. Math. Monthly 83(1976), 266-268; 87(1980), 292-294.

81. S. Reich, Extension problems for accretive sets in Banach spaces, J. Functional Analysis 26(1977), 378-395.

82. S. Reich, Almost convergence and nonlinear ergodic theorems, J. Approximation Theory 24(1978), 269-272.

83. S. Reich, Weak convergence theorems for nonexpansive mappings in Banach spaces, J. Math. Anal. Appl. 67(1979), 274-276.

84. S. Reich, Nonlinear ergodic theory in Banach spaces, Argonne National Laboratory Report #79-69, 1979.

85. S. Reich, Product formulas, nonlinear semigroups, and accretive operators, J. Functional Analysis 36(1980), 147-168.

86. S. Reich, Strong convergence theorems for resolvents of accretive operators in Banach spaces, J. Math. Anal. Appl. 75(1980), 287-292.

87. S. Reich, On the asymptotic behavior of nonlinear semigroups and the range of accretive operators I, II, Mathematics Research Center Report #2198, 1981; J. Math. Anal. Appl. 79(1981), 113-126; 87(1982), 134-146.

88. S. Reich, The almost fixed point property for nonexpansive mappings, Proc. Amer. Math. Soc., 88(1983), 44-46.

89. S. Reich, Averaged mappings in the Hilbert ball, J. Math. Anal. Appl., to appear.

90. J. Reinermann, G.H. Seifert, and V. Stallbohm, Two further applications of the Edelstein fixed point theorem to initial value problems of functional equations, Numer. Funct. Anal. Optimiz. 1(1979), 233-254.

91. R.C. Sine, On nonlinear contractions in sup norm spaces, Nonlinear Analysis 3(1979), 885-890.

92. P. Soardi, Existence of fixed points of nonexpansive mappings in certain Banach lattices, Proc. Amer. Math. Soc. 73(1979), 25-29.

93. D. Sullivan, The Dirichlet problem at infinity for a negatively curved manifold, preprint.

94. E. Thorp and R. Whitley, The strong maximum modulus theorem for analytic functions into a Banach space, Proc. Amer. Math. Soc. 18(1967), 640-646.

95. W.P. Thurston, Three dimensional manifolds, Kleinian groups, and hyperbolic geometry, Bull. Amer. Math. Soc. 6(1982), 357-381.

96. B. Turett, A dual view of a theorem of Baillon, Lecture Notes in Math., Vol. 80, Marcel Dekker, New York, 1982, pp. 279-286.

# INDEX

L



M



N



O



P